GENERAL, ORGANIC, AND BIOLOGICAL CHEMISTRY

A GUIDED INQUIRY

Second Edition

A Process Oriented Guided Inquiry Learning Course

MICHAEL GAROUTTE
Missouri Southern State University

ASHLEY MAHONEY
Bethel University

Kendall Hunt
publishing company

The POGIL Project
Director: Richard Moog
Associate Director: Marcy Dubroff

This book, including all interior line art was set by the authors.
Unless otherwise noted, source for all interior line art: Michael Garoutte and Ashley Mahoney
Cover image © Shutterstock.com

Kendall Hunt
publishing company

www.kendallhunt.com
Send all inquiries to:
4050 Westmark Drive
Dubuque, IA 52004-1840

To the instructor

In this second edition, the number of POGIL activities has been nearly doubled from the first edition, making it more appropriate for a variety of courses. Instructors may choose to use the majority of the activities in a **full-year GOB** (General-Organic-Biological) chemistry course sequence. Topics are flexible enough that appropriate sequences can be chosen for use in a **one-term GOB** course, a **one-term organic/biochem** course, or a **preparatory chemistry** course. Kendall-Hunt also allows custom publishing of subsets of the activities for a minimum number of copies. Each activity in this set may be used independently, but some questions may depend upon topics examined in earlier activities.

Materials are available to help instructors use this collection of POGIL activities effectively. These include suggested orders of activities for the types of courses listed above and facilitation guides for each activity. The facilitation guides give the content and process learning objectives, instructor facilitation notes, and suggested answers to the Critical Thinking Questions. Solutions to the Exercises at the end of each activity are also available. The facilitation guides are not for distribution to students, although the learning objectives and Exercise solutions may be shared with them. Please contact Danielle Schlictmann at Kendall Hunt Publishing (dschlichtmann@kendallhunt.com) for more information and access to these materials.

The Process Oriented Guided Inquiry Learning (POGIL) Project supports the dissemination and implementation of these types of materials for a variety of chemistry courses (high school, organic, physical, *etc.*) and in other disciplines such as materials science, biology and mathematics. Information about The POGIL Project and its activities, including additional materials, workshops, and other professional development opportunities, can be found at www.pogil.org.

Feedback regarding the effectiveness of the materials and suggestions for improvements would be appreciated. Send this assessment information to the authors by email (addresses available from our University websites).

How to use this book (for instructors)

How instructors use these materials varies, but during most POGIL sessions:

- **Students work in teams of three or four** to answer the *Critical Thinking Questions*. The Learning Cycle is used as the basis for a POGIL activity. Through a carefully crafted guided inquiry approach, the **exploration** of a model occurs through direct questioning. **Concept invention** then takes place as students begin to see patterns and relationships in the data and **terms are introduced.** Finally, questions are posed that ask students to address the **application** of a concept to new situations. See https://pogil.org/additional-resources for more information.

- **Each activity focuses on a particular process skill.** These process skills include both cognitive and affective processes that students use to acquire, interpret, and apply knowledge. At its outset, The POGIL Project identified seven process skills as those that would be the focus of development in a POGIL classroom. The skill targeted in each activity may be found in the instructor materials.

- **The instructor serves as the facilitator of learning** rather than the primary source of information. Effective facilitators spend significant class time observing student group work, asking and answering questions, leading whole-class discussions, and delivering just-in-time mini-lectures. More information is available at https://pogil.org/implementing-pogil.

- **Students are assigned specific roles,** a key feature of POGIL pedagogy.

- **These ChemActivities are designed to be a student's first introduction to a topic.** A very short review of the *previous* activity can be helpful when delivered at the beginning of the class. If you are employing a flipped classroom model or other structure that includes some formal lecture, make sure that any lecture on a topic comes *after* students have had the opportunity to discover that topic by completing the activity. If you find that students are having difficulty completing the activity in the allotted time, some instructors have achieved success getting a jump-start on class by asking students to complete the first few exploration questions of the activity on their own before coming to class.

- **Students will often ask you: "Is our answer right?"** Remind them that their job is to *construct a valid understanding of the underlying concepts*. Telling them the "right" answer can bring their processing of the ideas to a premature end. Ask how you can help. Usually students can rephrase their question to highlight the source of their confusion. Often, if they proceed to a following question, they will find the answer themselves.

- **These activities do not replace a traditional textbook, but rather enhance its use.** Any standard text may be used, and you are encouraged to correlate reading and/or homework assignments from the text for students to do *after* completing the ChemActivities in this book.

How to use this book (for students)

These activities are designed to be used by students working in teams during class. For each activity, read the Model or Information, and then work with your team to answer the Critical Thinking Questions (or "CTQs") that follow. For each CTQ, be sure to compare answers with your teammates before moving onto the next CTQ. Working together is everyone's collective responsibility. Studies show that this method of learning allows students to develop a deeper understanding of the material and retain it longer—even in later classes or on standardized exams.

In a class such as this, you may be frustrated at times because you cannot immediately see the "right" answer to a question. It is by design that some answers are not immediately obvious. Sometimes you will write an answer and go on to a later question, only to find that the later question causes you to reevaluate your earlier answer. This is OK! Later, when you have the "aha!" moment, you will not easily forget what you have learned.

If you are unsure of an answer, even after checking with your teammates, some good strategies are to read the following question, ask a nearby group, or (finally) pose a question to the instructor. However, try to avoid asking the instructor: "*Is our answer right?*" Instead, explain why you are confused, or ask a question that gets at the source of your confusion.

Before the next class, finish all assigned parts of the activity, including homework and reading from a textbook. Even better: find a study partner or group, and meet regularly! When you have study partners, you have a reason to be prepared (they are counting on you), and if you can't come up with the answer together, then you are less shy about asking the instructor.

Some comments from former students in this course

Being able to work with people was very beneficial. At first, it was a difficult adjustment to this learning style – but now I really enjoy it.

I thought that I wouldn't like this class but it has been really interesting to come to class every day and actually learn something that pertains to life.

If you work well with others and are able to learn as a group and be challenged by your group of peers to strive to really learn the subjects then I recommend this class. But if you'd rather work on your own and not get help from others then this class would be of no benefit to you.

Group discussions really encourage a better learning atmosphere that helps all group members understand the material better.

This is one of the hardest classes I have ever taken. But was the only one that taught me to seek out the answer instead of having it handed to you. This class will help me in future classes, because I have gained good study skills. For that I wish to thank [the instructor] :)

In high school I just memorized stuff, but now I finally understand.

Go to class if you don't want to do a whole lot else...even by showing up and doing the activities. I didn't have to spend a lot of time outside of class. I learned but didn't spend a huge amount of time doing it!

Acknowledgements

We would like to thank our numerous allied health students who have patiently worked through earlier versions of this material and have helped direct our revisions. Your questions and insights have taught us so much.

Michael P. Garoutte

I would like to thank several folks who have been influential in my path to author this book in the first edition and to work with Professor Mahoney on the second edition.

- To my graduate advisor, Richard L. Schowen, Professor Emeritus, University of Kansas: Dick, you said that I might want to consider authoring a textbook someday. I didn't forget. Thank you for the encouragement. It meant a lot, and still does.

- Andrei Straumanis and Renée Cole of The POGIL Project introduced me to POGIL at a workshop in 2003. They, and so many others in the project, have provided much advice and inspiration ever since. I can't imagine a more helpful and supportive group of folks.

- Rick Moog and Jim Spencer of Franklin & Marshall College didn't laugh when I showed them the first draft of these activities, and apparently believed in the adage "if you can't say something nice, don't say anything at all." Special thanks to Rick, who has continued to cheerfully answer any and all questions and has been all around encouraging in the entire process.

- My colleagues at Missouri Southern have supported my choice to "try something different" in class. Many teachers don't have it so good. Thanks especially to Professor Emeritus Mel Mosher, who gave me advice when I was new to teaching and continued to mentor me for the next decade and more. Mel passed away in 2010, and I still miss him.

- Thanks to the John Wiley & Sons editors Debbie Edson, Nick Ferrari, and Jennifer Yee, all of whom supported the publication of this work from its beginnings through the second edition. And thanks to The POGIL Project for continuing to support this work.

- Finally, I would like to dedicate my portion of this book to the people I share my life with—Susan, Audrey and Madeleine. Thank you for your ever-present love and support, especially during all the time I have spent working on this project. SDG

Ashley B. Mahoney

- To Rick Moog who encouraged me to jump in with both feet, and I haven't looked back.

- To many in the POGIL Project (Renée Cole, Andrei Straumanis, Rick Moog, Vicky Minderhout, Jenny Loertscher, and many others) who provided review of early drafts of the material and weren't afraid to be brutally honest. These activities are significantly better thanks to you.

- To the POGIL community for being welcoming and providing support and encouragement both in the classroom and in research. I have grown tremendously in both areas thanks to you. I also now have a well-stocked jewelry drawer!

- To my administrators and colleagues at Bethel University who supported me on many levels in trying a different approach and persevering through the initial trials. Thanks specifically to Rich Sherry and Ken Rohly who were willing to stand behind a nontenured professor who quit lecturing. Not many faculty have such tireless administrators on their side.

- To my close friends, extended, and immediate family for their patience and incredible support in this process. Thanks for keeping me grounded and reminding me of what is important in life. Hopefully meals will now improve (or not).
- To my parents for instilling a love of learning at an early age and providing me with a foundation built upon the rock.

Changes in the Second Edition

The collection of ChemActivities has been expanded from 44 to 76, and now includes topics typically encountered in a full year GOB sequence. The activities have improved, more robust models; specific prompts to encourage development of particular process skills, such as management, assessment, communication, teamwork, problem solving, information processing and critical thinking; and summary and reflector questions. An expanded Instructor's Manual available from the publisher contains Facilitation Guides for each activity listing the content and process objectives, suggestions for effective facilitation, and suggested responses to the Critical Thinking Questions for each activity.

If you are teaching a shorter GOB course, or even a preparatory chemistry course, subsets of these activities may fit the needs of your classroom. For example, the General Chemistry activities (CAs 1-32) are published separately under the title of *Introductory Chemistry: A Guided Inquiry*, and (for example) have been used in a preparatory chemistry course, along with Nivaldo Tro's *Introductory Chemistry*, 4th ed. Furthermore, the activities can be custom published in any amount or sequence to meet the needs of diverse courses by negotiation with the publisher.

Several of the ChemActivities are in multiple parts. Typically, this indicates that part A would be useful in a shorter GOB course, while the full set of activities would be appropriate for a full-year GOB course. The exception would be that in the Biochemistry section (CAs 44-56), the part A activities are typically a survey of the later parts, and would be skipped over entirely in a full-year sequence.

Table of Contents

Organic Chemistry Activities

Biological Chemistry Activities

Working in Teams; Estimation

Information: Brief description of roles

Much of the class time in this course will be spent working in teams of three or four. Each member of the team will be assigned to a particular role. Some typical roles (and their descriptions) are listed below. If a team member is absent on a particular day, then one member may have to fulfill more than one role. Your instructor will let you know how the roles will function in your course.

Manager (or Facilitator): Manages the team. Ensures that members are fulfilling their roles, that the assigned tasks are being accomplished on time, and that all members of the team participate together in activities and understand the concepts.

Spokesperson (or Presenter): Frequently the instructor will ask what the team responded to a particular question or whether the team agrees with another team's response. It is the spokesperson's role to reply to these questions. If an outside source is needed, the spokesperson ensures that everyone in the team agrees on what question to ask.

Quality Control: Guides the team to build consensus and ensures that the team agrees on responses to questions. Verifies that the team's answers to Critical Thinking Questions are consistent on paper. Ensures that all team members make revisions, if necessary, after class discussion.

Process Analyst (Reflector, Strategy Analyst): Observes and comments on team dynamics and behavior with respect to the learning process. Reports to the team periodically on how the team is functioning. For example, the Process Analyst might comment that a particular team member is dominating the discussion, or that the team needs to pause to allow one member to catch up.

Recorder: Records (on report form) the names of each of the team members at the beginning of each day. Keeps track of the team answers and explanations, along with any other important observations, insights, *etc.* The completed report with answers to any questions asked may be submitted to the instructor at the end of the class meeting.

Model 1: A centimeter ruler

Critical Thinking Question:

1. Estimate the number of table tennis ("ping-pong") balls that would completely fill the room you are working in. First, decide upon a "plan of attack" as a team. You may or may not choose to use the centimeter ruler. For the purposes of this exercise, you may choose to assume that the room is rectangular in shape and that it is completely empty of desks, people, *etc.* You may get up and move around the room. When your team has an answer, the spokesperson may be asked to write it on the board.

Exercise:

1. Read the assigned pages in the text, and work the assigned problems.

Types of Matter; Chemical and Physical Changes
(How can we classify matter?)

Model 1: Examples of some pure substances at room temperature

Item	Classification	State (or states)	Formula
aluminum	element	solid	$Al(s)$
hydrogen	element	gas	$H_2(g)$
mercury	element	liquid	$Hg(l)$
baking soda	compound	solid	$NaHCO_3(s)$
table salt	compound	solid	$NaCl(s)$
water	compound	liquid	$H_2O(l)$

Critical Thinking Questions:

Refer to Model 1 to help you answer Critical Thinking Questions (CTQs) 1-3.

1. What is the formula for hydrogen gas? _____ For liquid water? _____

2. How can you distinguish elements from compounds based on their chemical *formulas*? Consult with your team and write your consensus answer in a complete sentence or two.

3. Based on your answer to CTQ 2, complete the definition for

   A *compound* is composed of at least _____ different _____ that are combined chemically.

Model 2: Examples of some mixtures at room temperature

Item	Classification	State (or states)	Formula (or formulas)
hydrogen peroxide solution (3%)	homogeneous mixture	aqueous solution	$H_2O(l)$ and $H_2O_2(aq)$
salt water	homogeneous mixture	aqueous solution	$H_2O(l)$ and $NaCl(aq)$
coffee ("black")	homogeneous mixture	aqueous solution	$H_2O(l)$ and many others
muddy water	heterogeneous mixture	liquid + solid	$H_2O(l)$ and other stuff

Critical Thinking Questions:

4. The 3% hydrogen peroxide solution available in drugstores is 97% water. What is the formula for the hydrogen peroxide present in this solution (consult Model 2)?

5. Elements and compounds are considered pure *substances*. Compare Models 1 and 2. How does a *substance* differ from a *mixture*? Discuss with your team and write a consensus answer.

6. Devise a team hypothesis about the meaning of the labels (s), (l), (g), and (aq) on the formulas.

Information: States of matter

Matter can be classified by its physical **state** (or **phase**): **solid, liquid,** or **gas**. Most solids can be melted and even vaporized if the temperature is high enough.

The **phase labels** (s), (l), or (g) can be written after a formula to signify the physical state. So, $H_2O(g)$ would mean gaseous water, *i. e.,* water vapor.

Model 3: Equations for some chemical and physical changes

	Equation	Type of change
I	$H_2O(l) \rightarrow H_2O(s)$	physical
II	$2\ H_2(g) + O_2(g) \rightarrow 2\ H_2O(g)$	chemical
III	$2\ H_2O_2(aq) \rightarrow 2\ H_2O(l) + O_2(g)$	chemical
IV	$C_3H_6O(l) \rightarrow C_3H_6O(g)$	_____

Critical Thinking Questions:

*Questions 7-10 refer to Model 3. For each question, **manager:** ask a different team member to begin discussion by explaining his or her answer to the team.*

7. Write a complete sentence to describe in words (no formulas) what is happening in Equation I. Why is this process considered to be a *physical* change?

8. Describe in words what is happening in Equation II. Why is this a *chemical* change?

9. Describe in words what is happening in Equation III. Why is this a *chemical* change?

10. The formula for acetone is C_3H_6O. Without using formulas, write a sentence to describe describe what is happening in Equation IV. Is this a chemical or physical change? Fill in the blank in the Model.

Model 4: Flow chart for classifying matter

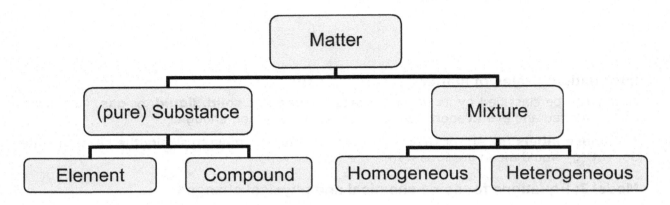

Information: Classifications of matter

Matter can be divided into two main types: pure **substances** and **mixtures** of substances.

A **substance** cannot be separated into other kinds of matter by physical processes such as filtering or evaporation, and is either an **element** (e. g., aluminum) or a **compound** (e. g., H_2O). Compounds are made of two or more elements chemically combined. The elements themselves cannot be separated into simpler substances even by a chemical reaction.

On the other hand, **mixtures** can be separated by physical means. Mixtures that have the same composition throughout are called **homogeneous** (e. g., salt water); those that do not are called **heterogeneous** (e. g., Italian salad dressing).

Critical Thinking Questions:

11. Have one team member reread your team's answer to CTQ 5 out loud. Add to or revise it if necessary.

12. As a team, try to brainstorm a physical method that you could use to separate salt water into two pure substances. Write your idea here.

Model 5: Some different representations of the water molecule

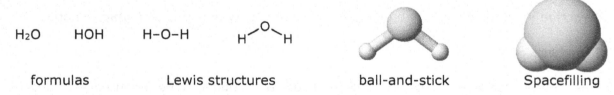

formulas	Lewis structures	ball-and-stick

Spacefilling

Critical Thinking Questions:

13. The "2" in H_2O is subscripted (written below the line). Based on Model 5, what do subscripts following an element in a formula represent?

14. Look at a periodic table of the elements. About how many elements are known? _____

15. Approximately how many elements are metals? (Estimate, don't count!) _____

16. There are **thousands** of organic compounds known—compounds formed out of only a few different elements (carbon, hydrogen, oxygen, nitrogen). How might this be possible?

17. Did everyone in your team contribute to the activity today? If so, explain how. If not, identify what individuals need to do to ensure participation by all in the next session.

Exercises:

1. Write the formula of each molecule for which the ball-and-stick structures are shown.

Key: ● = carbon   ○ = oxygen   ○ = hydrogen

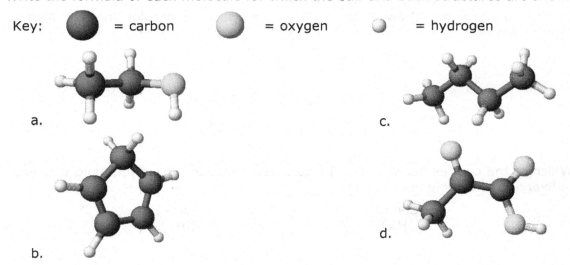

a.

b.

c.

d.

2. Using a periodic table, identify the elements represented in each formula, and state the number of atoms of each element in the formula. The first one has been done for you.

   a. NH_3 (ammonia)   *one nitrogen atom, three hydrogen atoms*

   b. $C_6H_{12}O_6$ (glucose)

   c. $Mg(OH)_2$ (milk of magnesia)

   d. H_2SO_4 (sulfuric acid, "battery acid")

   e. $C_{17}H_{18}F_3NO$ (fluoxetine, Prozac)

3. Using the flow chart in Model 4 to help you, first classify each of the following as either a **mixture** or pure **substance**. Then, for each **substance**, tell whether it is an **element** or a **compound**. For each **mixture**, tell whether it is **homogeneous** or **heterogeneous**; then list two or more components of the mixture.

   a. a lead weight

   b. apple juice

c. baking soda ($NaHCO_3$)

d. air

e. a 14-karat gold ring

f. a 24-karat gold coin

g. helium in a balloon

h. beach sand

i. concrete

j. whole blood

k. carbon dioxide

4. In the space below, draw a picture of three water molecules in the ball-and-stick representation.

5. Which of the choices below (I or II) would best represent the three molecules you drew in Exercise 4? Explain your choice.

Choice I	Choice II
H_6O_3	$3 H_2O$

6. Can you think of some commercial products you might have at home that are heterogeneous mixtures? List one or more.

7. Learn the **names** and **symbols** of the elements your instructor suggests. A good starting point is the first 30 elements, plus Br, Sr, Ag, Sn, I, Ba, Pt, Au, Hg, Pb. Spelling counts! You do **not** need to memorize any **numbers**, as a periodic table will always be available for your use.

8. Read the assigned pages in the text, and work the assigned problems.

Atoms and the Periodic Table[*]
(What are atoms?)

Model 1: Schematic diagrams for various atoms

● proton (+)
◉ neutron (no charge)
○ electron (-)

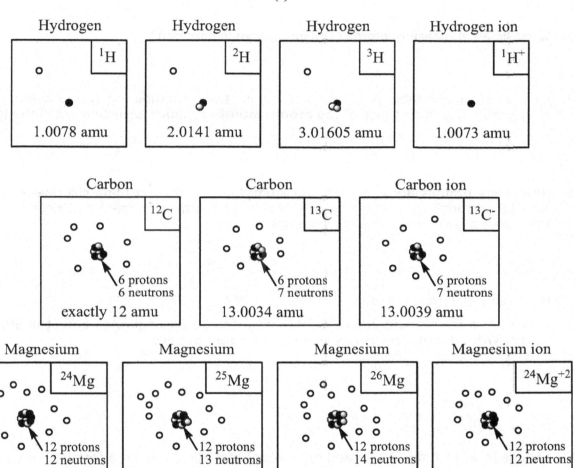

^1H, ^2H and ^3H are **isotopes** of hydrogen. ^{12}C and ^{13}C are **isotopes** of carbon.

^{12}C and ^{13}C may also be written as "carbon-12" and "carbon-13"

The **nucleus** of an atom contains the protons and the neutrons (if any).

An "amu" is an **atomic mass unit**.

---

[*] Adapted from ChemActivity 1, Moog, R.S.; Farrell, J.J. *Chemistry: A Guided Inquiry*, 5th ed., Wiley, 2011, pp. 2-6.

Critical Thinking Questions:

Manager: *For questions 1 – 3, identify a different team member to give the first explanation.*

1. Look at the schematic diagrams for carbon. What do all three carbon atoms (and ions) have in common?

2. What do all four hydrogen atoms (and ions) have in common?

3. What do all magnesium atoms (and ions) have in common?

4. Look at a periodic table (for example, in your text). Considering your answers to CTQs 1-3, what is the significance of the **atomic number**, above each element in the table?

5. How many protons are in all chlorine atoms?_____ Do you think chlorine atoms exist with 18 protons? _____ Why or why not? Confer with your team and once you agree, write your answer in a complete sentence.

6. How many electrons are found in ^{12}C? _____ ^{13}C? _____ $^{13}C^-$? _____

7. Engage in a team discussion to identify what feature distinguishes a neutral atom from an ion. Write the consensus answer in a complete sentence.

8. Engage in a team discussion and once your team agrees, write a formula for calculating the charge of an ion. ***Spokesperson (Presenter):*** be prepared to share the team answer with the class.

9. A positively charged ion is called a cation (pronounced "cat ion"), and a negatively charged ion is called an anion (pronounced "an ion"). Which term applies to the magnesium ion?

10. ***Process Analyst (Reflector):*** comment to your team on strengths and needed improvements for your teamwork so far.

11. a. Calculate the charge for a chlorine (Cl) ion if a chlorine atom gains an electron to become an ion.

    b. Is this chlorine ion a **cation** or **anion** (circle one)?

12. In a box in the corner of each schematic diagram in Model 1 is the element symbol and the *mass number* for the atom (superscript on the left side of the element symbol).
    a. How is this mass number determined? Write your team answer as a <u>formula</u>.

    b. As a team, propose a reason why is it called a "mass" number.

13. Where is the majority of the mass of an atom located? Write your team answer.

14. What is the mass number for the following atoms:
    a) carbon-12 _____   b) ^{13}C _____   c) ^{37}Cl _____   d) uranium-238 _____

15. What structural element do all isotopes of an element have in common? How are their structures different? Write the consensus answers.

16. Considering what you know about isotopes, do all atoms of an element weigh the same? _____(yes/no). Explain in a sentence. **Spokesperson:** be prepared to share the team answer with the class.

17. As a team, identify three major concepts developed from Model 1, and write them below.

18. Identify one unanswered question that remains within your team concerning Model 1.

Model 2: The periodic table

Look at a periodic table (such as in your textbook).

There are 18 columns, known as **families** or **groups**. The groups can be numbered in various ways. We will use the numbers 1-18 for group designations. Elements in the same group often have similar properties. Some groups have family names.

Your table may be divided into the metals, semimetals, and nonmetals. If not, note that a "stair-step" line which starts between boron (B) and aluminum (Al) roughly divides metals from nonmetals. Semimetals (or "metalloids") lie along the line.

Critical Thinking Questions:

Manager: *Identify a different team member to offer the first answer for questions 19-21.*

19. From your experience, list some properties of metals.

20. Locate on your periodic table which group of elements represents metals.

21. From your experience, list some properties of nonmetals.

22. Locate on your periodic table which group of elements represents non-metals.

23. Are most elements metals or nonmetals? _____

24. Transistors and computer "chips" are made of semiconductor materials. What kind of elements would be used for this purpose? Give an example.

25. **Process Analyst (Reflector):** how did your team members perform their roles today? How might you perform better next time? Write the suggestions for improvement.

Exercises:

1. Complete the following table.

Symbol	Atomic Number	Mass Number	Number of Protons	Number of Neutrons	Number of Electrons
^{40}K					
			15	17	18
Zn^{2+}				38	
	35	81			36

2. Complete the table below, classifying each element as a metal, nonmetal, or semimetal. Watch spelling!

Symbol	Name	Classification
Pd		
Cl		
	germanium	

3. Complete the table below, classifying each element as an alkali metal, alkaline earth metal, transition metal, halogen, or noble gas. Watch spelling!

Symbol	Name	Classification
F		
Ca		
	rubidium	
Cr		
	krypton	

4. Learn the family names of groups 1, 2, 17, 18.

5. Learn to recognize which elements are metals, nonmetals, semimetals (metalloids), transition elements or main group elements.

6. Read the assigned pages in the text, and work the assigned problems.

Unit Conversions: Metric System
(What is a conversion factor?)

Model 1: Fuel efficiency of a particular automobile

A particular automobile can travel 27 miles per gallon of gasoline used.

The automobile has a 12-gallon gasoline tank.

At a particular location, gasoline costs $4.00 (4.00 USD) per gallon.

Critical Thinking Questions:

1. Three statistics are given in Model 1. Circle the two statements that give numerical **ratios**.

2. One statement in Model 1 gives a **measured quantity**. Write the quantity (with the associated **unit**).

3. Write each ratio that you circled in Model 1 as a **fraction**. Your fraction should have a **number** and a **unit** in both the numerator and the denominator of the fraction. Check that all team members have the same two fractions.

4. How many miles can the automobile travel on a full tank of gasoline? Show your work by writing the **quantity** from CTQ 2 multiplied by the appropriate **fraction** from CTQ 3. Show all units.

5. Discuss with your team and reach a consensus why the **answer** to CTQ 4 does not include the unit "gallons."

6. Explain why the **fraction** used in CTQ 4 may be called a **conversion factor**.

7. Do all four conversion factors below give equivalent information? Discuss with your team and write a consensus explanation.

$$\frac{27\ \text{mi}}{1\ \text{gal}} \qquad \frac{27\ \text{mi}}{\text{gal}} \qquad \frac{1\ \text{gal}}{27\ \text{mi}} \qquad \frac{\frac{1}{27}\ \text{gal}}{\text{mi}}$$

Information:

When reading a conversion factor out loud, the line is pronounced "per." So, the first conversion factor in CTQ 7 would be read "twenty-seven miles per one gallon." Since the other ratios (including "one-twenty-seventh gallon per mile") are either the same or the reciprocal, all ratios give the same information.

Model 2: Definitions of the inch and the foot

1 inch = 2.54 cm (exactly)

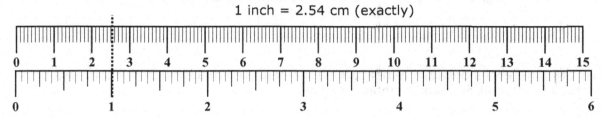

There are exactly 12 inches in one foot.

Critical Thinking Questions:

8. How many centimeters are in one inch?

9. Draw a large X through each **incorrect** conversion factor below.

$$\frac{1\ cm}{2.54\ in} \qquad \frac{2.54\ cm}{1\ in} \qquad \frac{2.54\ in}{1\ cm} \qquad \frac{1\ in}{2.54\ cm}$$

10. Suppose you want to convert a person's height from inches into centimeters. Circle the conversion factor in CTQ 9 that you would use. Explain why your choice will eliminate the units of "inches" from the height.

Model 3: The unit plan

A unit plan begins with the **unit of the known or measured quantity** and shows how the units will change after multiplying by each conversion factor used, in order. The unit plan for CTQ 10 would be:

$$in \rightarrow cm$$

Each **arrow** in the unit plan represents **one conversion factor**.

Critical Thinking Questions:

11. A basketball player is seven feet tall.

   a. Using Models 2 and 3 for reference, place **units** in the boxes to complete the unit plan for converting the height of the basketball player into centimeters.

$$feet \rightarrow \boxed{} \rightarrow \boxed{}$$

   b. Write one conversion factor (as in CTQ 9) corresponding to **each** of the two arrows in part (a). Include all units.

c. Perform the calculation by multiplying the **measured quantity** (in this case, the number of feet) by each **conversion factor** in order. Show all work with units.

d. Why did you need to convert first from feet into inches even though you wanted to know the height in centimeters? Write a sentence to explain.

12. List two differences between a measurement (measured quantity) and a conversion factor. Ensure that all team members agree.

Model 4: Conversion factors from metric system prefixes

Metric Base Units

Distance: meter (m)
Mass: gram (g)
Volume: liter (L)
Time: seconds (s)

Table 1: Some metric system prefixes

Prefix	Symbol	Factor	Meaning
mega	M	10^6	one million
kilo	k	10^3	one thousand
deci	d	10^{-1}	one tenth
centi	c	10^{-2}	one hundredth
milli	m	10^{-3}	one thousandth
micro	μ	10^{-6}	one millionth

The metric system prefixes can be associated with any unit. So, for example, one can write millimeters (mm), milligrams (mg), or milliliters (mL).

In the metric system, the **base unit** is the unit without prefixes. When converting units within the metric system, it is useful to convert into the base unit first. The **factors** listed in Table 1 give you an **equality** to the base unit. For example:

   $1 \text{ mm} = 10^{-3} \text{ m}$ or $1 \text{ mm} = 0.001 \text{ m}$ $1 \text{ kg} = 10^3 \text{ g}$ or $1 \text{ kg} = 1000 \text{ g}$

The equality can be used to create a **conversion factor**. For millimeters, the possible conversion factors are[*]:

$$\frac{1 \text{mm}}{10^{-3} \text{m}} \text{ and } \frac{10^{-3} \text{ m}}{1 \text{mm}}$$

Critical Thinking Questions:

13. How many mm are in 1.89 m? In the setup below, add a conversion factor from those shown in Model 4 so that the units of meters will cancel out and leave millimeters. Then calculate the answer.

> To enter 10^{-3} into most scientific calculators, enter the number **1** followed by the **exponent** (EE or EXP or 10^x) key and then **−3** (or 3−). Do not use the **power** (y^x) key.

   $1.89 \text{ m} \times \underline{\hspace{3cm}} =$

_____

[*] If you have already memorized other equivalent conversions such "1000 mm = 1 m," you may use them. However, this book will always use the powers of 10 and prefixes shown in Table 1.

14. In question 13, what would the final **units** of the answer be if the conversion between millimeters and meters had been written upside down ($\frac{10^{-3}\,m}{1mm}$)? Check that all team members understand this, and why it is incorrect.

15. How many centimeters are in 2.2045×10^{-2} kilometers?
    a. Complete the unit plan. (Remember to convert to the base unit first.)

    $$km \rightarrow \qquad \rightarrow$$

    b. Each arrow in part (a) requires a conversion factor. Write the two conversion factors needed for part (a). Ensure all team members agree.

    c. Perform the calculation, showing all work.

16. Provide one grammatically correct sentence that describes the most important unanswered question that your team has about today's activity.

17. List two problem-solving strategies or insights your team discovered today.

Exercises:
1. Write unit plans, and then perform the unit conversions below. Use only the conversion factors given in the problem or those given in Model 4.
    a. Convert 36 mg to g.

    Complete the unit plan: mg $\rightarrow$ Complete the equality: 1 mg =

    Write the two possible conversion factors:

    Choosing the correct conversion factor, perform the conversion:

b. Convert 5.51 ms to µs.

Unit plan:

Conversion factors:

Perform the conversion:

2. A traveler wants to know the cost (in dollars) of the gasoline to travel 75 miles in the automobile in Model 1.

   a. Write a unit plan for the conversion. (Remember to begin with the known quantity).

   b. Write the two conversion factors needed for the conversion.

   c. Perform the calculation, showing all work and units.

3. What is the weight in grams of a 7.5-karat diamond engagement ring? There are exactly 200 mg in one karat. Make a unit plan first.

4. There are 5280 feet in a mile. Convert 161 km to miles, using only equalities taken from this activity. Make a unit plan first.

Information: Units of volume

One *milliliter* (mL) is equal to one *cubic centimeter* (1 cc or 1 cm^3).

One *liter* (L) is equal to one *cubic decimeter* (1 dm^3).

5. Convert 1000 cm³ to dm³ using only the equalities given in the information above and the metric system prefix *milli* (10^{-3}). Show work. Make a unit plan first.

6. Measurements may have complex units. For example, the speed of light is 299,792,458 m/s. What is this speed in km/hour?

   It is helpful to write out the measurement to clearly show the numerator and denominator, as follows:

   $$\frac{299,792,458\,\text{m}}{1\,\text{s}}$$

   One way to solve this problem is to make two unit plans: one for the numerator and one for the denominator.

   a. The unit plan for the numerator is m → km. Complete the unit plan for the denominator: s →   →

   b. Write the three conversion factors needed.

   c. Perform the calculation for the numerator.

   d. Perform the calculation for the denominator.

   e. Divide the numerator by the denominator to get the final answer, including units.

7. How long does it take for light to travel 100 meters? (Unit plan: m → ?)

8. The daily dose of ampicillin for the treatment of an ear infection is 115 mg ampicillin per kilogram of body weight (115 mg/kg). What is the daily dose of the drug for a 27-lb child? Make a unit plan first.

9. Read the assigned pages in the text, and work the assigned problems.

Measurements and Significant Figures
(How do we measure things scientifically?)

Model 1: Outlines of a coin (U.S. quarter dollar) and a glass bottle on top of a centimeter ruler

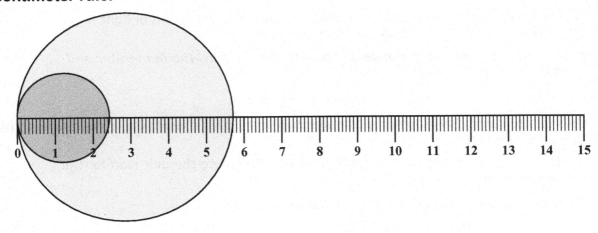

Critical Thinking Questions:

1. Use the ruler shown to determine the diameter of the quarter in Model 1 in centimeters. Express your answer as a **decimal number with 2 digits following the decimal point**. Include units.

2. How did you determine the final decimal place in the measurement in CTQ 1? Discuss with your team and once your team agrees, write your answer in a complete sentence.

3. Use the ruler shown to determine the diameter of the glass bottle in Model 1 in centimeters. Express your answer as a decimal number with two digits after the decimal point. Include units.

4. Explain how you determined the final decimal place in the measurement in CTQ 3.

5. Discuss for half a minute or so with your team why it would make little sense to write your measurement in CTQ 3 with **three** decimal places. Record your team's top reason.

6. Write the two conversion factors that relate inches to centimeters.

7. Convert the diameter of the glass bottle to inches. Show work. Compare your answer with your team members.

Information: Precision

Generally, when measurements are reported, there is **uncertainty in only the final digit**. This leads to a problem deciding how to report the answer to CTQ 7. The measured diameter of the bottle has (at most) three digits of **precision** (for example, 5.70 cm) to begin with. These three digits are called **significant figures** ("sig figs"). When the calculation to convert to inches is performed, there are many digits after the decimal point. One cannot **create** more precision just by changing to the English system, so the result should be **rounded off** to three sig figs. If the digit following the third significant figure is at least 5, round **up**; otherwise, simply drop the trailing digits.

Critical Thinking Questions:

8. 2.244094488 cm rounded to three "sig figs" becomes _____

9. 2.248031496 cm rounded to three "sig figs" becomes _____

10. According to the U.S. mint, quarters are actually 2.426 cm in diameter. Considering significant figures, does your measurement in CTQ 1 agree with this? Explain. Confer with your team and once your team agrees, write your answer in a complete sentence.

11. Convert the official diameter of a quarter into inches. Show your work. Round your answer to the correct number of significant figures.

12. As a team, reflect on your work so far. What is one way in which you are working well together as a team, and what is one way you can improve? Outline a strategy for improvement for the remainder of the activity.

Information: Determining whether zeroes in reported numbers are significant

Table 1: Rules for significant figures in measurements

Rule	Example (sig figs are underlined)
1. All non-zeroes are significant	<u>14</u>, <u>2.25</u>, <u>1999</u>
2. <u>Leading</u> zeroes are **not** significant	0.<u>125</u>, 0.00<u>54</u>, 0.0<u>5</u>
3. Zeroes <u>trapped</u> between nonzero digits **are** significant	<u>409</u>, 0.0<u>302</u>, <u>2002</u>
4. <u>Trailing</u> zeroes are significant **only** if an explicit decimal point is present*	<u>7</u>00, <u>700.</u>, <u>7.00</u>×10²

Critical Thinking Questions:

13. What is a *leading zero*? Can they be before the decimal, after, both or neither? Discuss with your team, and write a definition of *leading zero*.

14. Give the number of significant figures in each measurement. Then state which rule from Table 1 you used to help you decide your team's answer.

    a. 100 m _____   Rule # ____

    b. 0.00095 g _____   Rule # ____

    c. 3600.0 s _____   Rule # ____

15. Write the following measurements in scientific notation, without changing the number of significant digits.

    a. 100 m _____

    b. 0.00095 g _____

    c. 3600.0 s _____

16. Explain why 3.6×10^3 is *not* the answer to CTQ 15c. Write your team answer.

17. Round each of the following to three significant digits.

    a. 0.00320700 L _____

    b. 3.265×10^{-4} m _____

    c. 129762 s _____

18. List any questions that remain within your team about significant figures.

---

* See why scientific notation is useful? It eliminates any ambiguity.

Information: Significant figures in calculations

When converting units, numbers that are **exact** by definition do not affect the number of significant figures in the answer. This includes conversions within a measurement system (and also 1 in = 2.54 cm, such as in CTQ 7).

Table 2 gives some rules of thumb that help determine how much precision to report when doing calculations. These rules involve determining which digits in each number are **significant** (i. e., they indicate the precision) and how to treat them in calculations.

Table 2: Rules for calculations with significant figures[†]

Operation	Rule	Example (sig figs underlined)
Multiplication and division	Keep the smallest number of **sig figs**.	$250 \div 7.134 = 35.043$ $= \mathbf{35}$
Addition and subtraction (do **after** multiplication and division)	Keep the smallest number of **decimal places**.	73.147 52.1 + 0.05411 = 125.30111 = **125.3**
Logarithms	Only digits in the mantissa (after the decimal) are significant	$\log(2003) = 3.3016809$ $= \mathbf{3.3017}$

Since most calculations in chemistry involve multiplication or division, we can often use the following rule of thumb: Keep only as many significant figures in your answer as in the measurement with the fewest number of sig figs.

Also, to avoid compounding errors, one should never round off intermediate results, but wait to round until you have the final answer.

19. Indicate the number of significant figures the results of the following calculations should have. Then solve the problem, rounding to the correct number of digits (optional).

   a. 503.2×1.0 # of sig figs: ____   solution: _____

   b. $151.00 + 1.042 + 3061$ # of sig figs: ____   solution: _____

   c. $\log(2.6 \times 10^{-3})$ # of sig figs: ____   solution: _____

   d. $40.0 \times 6203 \div 22000$ # of sig figs: ____   solution: _____

20. Did everyone in your team understand the material in the activity today? If so, explain how your team ensured that everyone understood. If not, identify what your team needs to do to assure that everyone in the team understands the material in the next session.

[†] These rules are simple to apply but only provide an **estimate** of the correct level of precision. This estimate is preferable to simply including all possible digits in the result of a calculation, but not as good as a rigorous error analysis.

Exercises:

1. Explain why measurements have uncertainty.

2. Consider the quarter in Model 1 and its diameter that you measured in centimeters.
   a. Write a unit plan (no numbers) to convert the diameter into millimeters.

   b. Write the two conversion factors needed.

   c. Perform the conversion, showing all work.

3. Follow the procedure in Exercise 1 to convert the diameter of the glass bottle into **micrometers**.

4. Perform the calculations, reporting your answer with the correct number of sig figs.
   a. 26.234 g – 5.6 g = _____

   b. 67.6 oz ÷ 8.0 oz/cup = _____cups

   c. $189 \text{ cm} + 6.0 \text{ in} \times \dfrac{2.54 \text{ cm}}{1 \text{ in}}$ = _____cm

5. Calculate the circumference of the glass bottle from Model 1 in centimeters. Report your answer with the correct number of significant figures. (Circumference = π × diameter.)

6. Consider the picture of the partially filled burette at the right. Circle the reading below that shows the correct level of precision. Explain your answer.

   15 mL   15.4 mL   15.40 mL   15.400 mL

7. Read the assigned pages in your text, and work the assigned problems.

Density and Temperature
(What is measurement useful for?)

Critical Thinking Question:
1. Which weighs more, a ton of bricks or a ton of cotton balls?

Information
CTQ 1 is a "trick" question sometimes found in children's joke books. Of course, the bricks and the cotton balls weigh the same (1 ton). But the cotton balls take up more space (volume), making them less **dense**. *Density* is a measure of the mass of a particular volume of something.

$$\text{density} = \frac{\text{mass}}{\text{volume}} \quad or \quad d = \frac{m}{V}$$

Critical Thinking Question:
2. A piece of tin metal weighing 85.251 g is placed into 41.1 cm^3 of ethyl alcohol (d = 0.798 g/cm^3) in a graduated cylinder, and it sinks to the bottom. The alcohol level increases to 52.8 cm^3.

    a. What is the mass of the tin?

    b. What is the volume of the tin?

    c. Write the volume of the tin in words. (Hint: what does cm^3 mean?)

    d. What is the density of the tin? Show all work, including units. Compare with your teammates to make sure that your answers are exactly the same.

Information
Sometimes, something called **specific gravity** is used to report density. The specific gravity (sp gr) is a ratio of the density of something to the density of water.

$$\text{specific gravity (sp gr)} = \frac{\text{density of sample}}{\text{density of water}}$$

Since the density of water at room temperature is 1.00 g/mL, the sp gr of a sample is the same as the density, except that no units are given. So, <u>if you know the sp gr, just add the units of g/mL to give the density</u>.

Critical Thinking Questions:
3. What is the specific gravity of the tin in CTQ 2?

4. If you have 750 mL of 190 proof Everclear (95.6% ethyl alcohol, d = 0.80 g/mL), how many grams of liquid do you have? (Hint: first, use algebra to rearrange the density equation to solve for *mass*.)

Model: Three temperature scales, Fahrenheit, Celsius and Kelvin

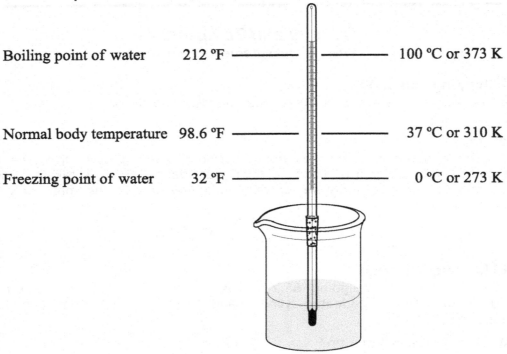

Boiling point of water   212 °F ————————— 100 °C or 373 K

Normal body temperature   98.6 °F ————————— 37 °C or 310 K

Freezing point of water   32 °F ————————— 0 °C or 273 K

Normal body temperature is 98.6 <u>degrees Fahrenheit</u> (98.6 °F)
There are 66.6 <u>Fahrenheit degrees</u> (F°) between the freezing point of
water and body temperature.

Critical Thinking Questions:

5. Write an equation to convert Celsius and Kelvins.

    K = °C + _____

6. How many Celsius degrees are there between the freezing and boiling points of water?_____

7. How many Kelvins are there between the freezing and boiling points of water? _____

8. Considering your answers to CTQs 6 and 7, how is the size of **one** Celsius degree related to the size of one Kelvin (circle one): The °C is   **larger**   **the same size**   **smaller** Discuss with your team and when you reach agreement, write an explanation of your answer.

9. What is the boiling point of water in Fahrenheit? _____°F

10. How many Fahrenheit degrees are there between the freezing and boiling points of water?_____

11. Compare your answers to CTQs 6 and 10. Fill in numbers in the numerators and denominators to complete the two **ratios** (conversion factors) below.

$$\frac{F°}{C°} \qquad\qquad\qquad \frac{C°}{F°}$$

12. The freezing point of water is the reference (zero) point for the Celsius scale. At the freezing point of water, what number of **Fahrenheit** degrees must be subtracted from the temperature so that it will also be the **zero point**?

13. So, to convert from °F to °C, two adjustments are needed. One has to subtract 32 to get to the same reference point (0 °C), and then use a conversion factor (see CTQ 11). Using only your answers to CTQ 11 and 12, complete the equation below.

$$°C = (°F - \qquad) \times \left(\frac{C°}{F°} \right)$$

14. To convert °C to °F, one has to use the conversion factor to convert to °F first, and then add 32 °F units ("adding apples to apples"). Write an equation for this.

$$°F =$$

15. Some textbooks give the ratio 9/5 as the conversion factor between °F and °C. As a team, develop an explanation of how this is the same as the one in the equation you developed in CTQ 14.

16. Write grammatically correct English sentences to describe two important concepts your team has learned in class today.

17. Provide one grammatically correct sentence that describes the most important unanswered question about today's activity that remains with your team.

Exercises:

1. Suppose the room temperature is 72°F. Report this temperature in degrees Celsius and in kelvins.

2. The sp gr of mercury is 13.6. How many mL of mercury are in a barometer containing 2040 g of mercury? What is this volume in cm^3 (Hint: $1\ cm^3 = 1\ mL$)?

3. Your long lost great aunt from France left you a fancy French oven in her will. Unfortunately, it is calibrated in degrees Celsius, and you need to bake a cake at 350°F. At what temperature do you set the oven?

4. How many pounds of Everclear are contained in a 750-mL bottle (compare CTQ 4)?

5. Read the assigned pages in your text, and work the assigned problems.

Atomic Number and Atomic Mass[*]
(Are all atoms of an element the same?)

Model 1: Natural abundances of isotopes

Each element found in nature occurs as a mixture of isotopes. The isotopic abundance can vary appreciably from place to place in the universe. For example, the isotopic abundances in the Sun are much different than on the Earth. On Earth, however, the abundance shows little variation from place to place.

Table 1: Natural isotopic abundance on Earth for some elements.

Isotope	Natural Abundance on Earth (%)	Atomic Mass (amu)
^1H	99.985	1.01
^2H	0.015	2.01
^{12}C	98.89	12.00
^{13}C	1.11	13.00
^{16}O	99.76	16.00
^{17}O	0.04	17.00
^{18}O	0.20	18.00
^{24}Mg	78.99	23.99
^{25}Mg	10.00	24.99
^{26}Mg	11.01	25.98

$1 \text{ amu} = 1.66 \times 10^{-24} \text{ g}$

Critical Thinking Questions:

1. How many isotopes of hydrogen occur naturally on Earth? _____

2. How many isotopes of oxygen occur naturally on Earth? _____

3. What do all isotopes of oxygen have in common (recall ChemActivity 3)? How are they different? Check that all team members agree.

4. If you select one carbon atom at random, the mass of that atom is most likely to be _____ amu.

5. What is the mass (in amu) of 100 ^{12}C atoms? _____ Of 100 ^{13}C atoms?_____

---

[*] Adapted from ChemActivity 2, Moog, R.S.; Farrell, J.J. *Chemistry: A Guided Inquiry*, 5th ed., Wiley, 2011, pp. 8-12.

CA07

6. If you select one hundred carbon atoms at random, the total mass will likely be ____.

    a. 1200.0 amu

    b. slightly more than 1200.0 amu

    c. slightly less than 1200.0 amu

    d. 1300.0 amu

    e. slightly less than 1300.0 amu

Discuss your selection as a team, and explain your reasoning. Write the team summary in a complete sentence.

Model 2: The average mass of a marble

In a collection of marbles, 25% of the marbles have a mass of 5.00 g and 75% of the marbles have a mass of 7.00 g. The average mass of a marble is 6.50 g.

The average mass of a marble can be determined by dividing the total mass of the marbles by the total number of marbles:

$$\text{Average mass of a marble} = \frac{(1 \times 5.00) + (3 \times 7.00)}{4} = 6.50\,\text{g} \qquad (1)$$

The average mass of a marble can also be determined by first multiplying the fraction of marbles of a particular type by the mass of a marble of that type and then taking a sum over all types of marbles:

$$\text{Average mass of a marble} = (0.2500 \times 5.00\,\text{g}) + (0.7500 \times 7.00\,\text{g}) = 6.50\,\text{g} \qquad (2)$$

Critical Thinking Questions:

7. Do any of the marbles in Model 2 have the average mass of 6.50 g? _____

8. a. Working with your team, use the method of equation (2) in Model 2 to calculate the average mass of a carbon atom in amu. Show your work.

    b. Does any one carbon atom have this mass? Explain.

9. For any large collection of carbon atoms (randomly selected):

    a. What is the average atomic mass of a carbon atom (in amu) (CTQ 8a)?

b. What is this average mass in grams (1 amu = 1.66×10^{-24}g)?

10. What is the mass in grams of 6.022×10^{23} carbon atoms (randomly selected)?

11. a. Work with your team and use the method of equation (2) in Model 2 to calculate the average mass of a magnesium atom in amu.

b. Does any one magnesium atom have this mass? Explain.

For Questions 12 – 14, the manager should identify a different team member to offer the first explanation.

12. For any large collection of magnesium atoms (randomly selected):

a. What is the average atomic mass of a magnesium atom (in amu)?

b. What is this average mass in grams (1 amu = 1.66×10^{-24}g)?

13. What is the mass (grams) of 6.022×10^{23} magnesium atoms (randomly selected)?

14. Examine a periodic table and find the symbol for magnesium.

   a. How does the number given just below the symbol for magnesium (rounded to 0.01) compare with the average mass (amu) of one magnesium atom?

   b. How does the number given just below the symbol for magnesium (rounded to 0.01) compare with the mass (grams) of 6.022×10^{23} magnesium atoms?

15. As a team, give two interpretations of the number "24.305" found below the symbol for magnesium on the periodic table. Write the team's consensus answer in a complete sentence.

16. Does <u>any</u> magnesium atom have a mass of 24.305 amu? Does <u>any</u> chlorine atom have a mass of 35.453 amu? Discuss with your team and write a consensus explanation.

17. List the major concept(s) of this activity.

18. Identify two strengths of your teamwork today, and explain why these strengths are important.

Exercises:

1. Define isotope.

2. Isotopic abundances are different in other parts of the universe. Suppose that on planet Krypton we find the following stable isotopes and abundances for boron:

   ^{10}B (10.013 amu)   65.75%
   ^{11}B (11.009 amu)   25.55%
   ^{12}B (12.014 amu)   8.70%

   What is the value of the average atomic mass of boron on planet Krypton?

3. Indicate whether each of the following statements is true or false and explain your reasoning.

   On average, one Li atom weighs 6.941 amu.

   Every H atom weighs 1.008 amu.

4. What is the average mass, in grams, of one helium atom? Of one potassium atom?

5. The entry in the periodic table for chlorine contains the symbol Cl and two numbers: 17 and 35.453. Give two pieces of information for each number about the element chlorine.

6. Write a sentence or two to explain the difference between the mass number of a particular atom of an element and the average atomic mass of an element.

7. Read the assigned pages in your text, and work the assigned problems.

Nuclear Chemistry
(What is radiation?)

Information:

Some elements have isotopes with unstable nuclei and are considered **radioactive**. This means that their nuclei will break down ("decay") while giving off particles, "rays," or both. When an atom emits (loses) a radioactive particle, the atom changes into a new element.

Model 1: A nuclear reaction. Uranium-238 decays to produce thorium-234 and a helium nucleus.

$$\underset{92}{^{238}}U \rightarrow \underset{90}{^{234}}Th + \underset{2}{^{4}}He$$

Critical Thinking Questions:

1. Look at a periodic table and find the atomic number of uranium. In Model 1, what does the subscripted (bottom) number in front of uranium represent?

2. What does the superscripted (top) number in front of uranium (U) represent?

3. How many protons and neutrons are found in uranium in Model 1?
   _____ protons   _____ neutrons

4. How many protons and neutrons are found in thorium (Th) in Model 1?
   _____ protons _____ neutrons

5. What is the difference in:
   a. the number of protons between U and Th? _____
   b. the number of neutrons? _____
   c. the mass number? _____

Model 2: Nuclear reactions and ionizing radiation.

A **nuclear reaction** is a change in the composition of the nucleus of an atom. This is not normally considered a chemical reaction, and does not depend on what molecule the atom might be in.

Table 1: Some types of ionizing radiation produced in nuclear reactions

Type of Radiation	Symbol	Mass Number	Charge	Relative penetrating ability	Shielding required	Biological hazard
Alpha particle	α or $^{4}_{2}He$	4	2+	very low	clothing	none unless inhaled
Beta particle	β or $^{0}_{-1}e$	0	1−	low	heavy cloth, plastic	mainly to eyes, skin
Gamma ray	γ or $^{0}_{0}\gamma$	0	0	very high	lead or concrete	whole body
Neutron	$^{1}_{0}n$	1	0	very high	water, lead	whole body
Positron	β^{+} or $^{0}_{1}e$	0	1+	low	heavy cloth, plastic	mainly to eyes, skin

Critical Thinking Questions:

6. Look at the subscripts on the **symbols** in Table 1. Which column in the table has values that match these subscripts?

7. What does the superscript indicate in the **symbols** in Table 1?

8. In CTQ 1, you likely said that the subscripts indicated the number of protons in the nucleus, and in CTQ 6, you likely noticed that the subscripts indicate the charge. Which of these answers works in both Model 1 and Model 2? Work with your team to explain.

9. Consider the nuclear reaction from Model 1: $^{238}_{92}U \rightarrow ^{234}_{90}Th + ^{4}_{2}He$

   a. How does the number of protons in the reactant (on the left side of the arrow) compare with the total number of protons in the products (to the right of the arrow)?

   b. How does the mass number of the reactant compare with the total of the mass numbers of the products?

   c. How does the number of neutrons in the reactant compare with the total number of neutrons in the products?

   d. Based on Table 1, what type of radiation is produced in this reaction?

Information:

In nuclear reactions, both the total number of protons and the total number of neutrons remain unchanged. Another way of saying this is that the sum of the atomic numbers (and mass numbers) in the reactants must equal the sum of the atomic numbers (and mass numbers) in the products. You verified this in CTQ 9.

Critical Thinking Question:

10. Suppose that a gamma ray were also released in the reaction in CTQ 9. Discuss with your team how each side of the reaction equation would change. When you agree, write your team explanation in a complete sentence or two.

Information:

Table 2. Types of emission for some common radioisotopes

Radioisotope	Symbol	Radiation type
barium-131	$^{131}_{56}Ba$	γ
carbon-14	$^{14}_{6}C$	β
chromium-51	$^{51}_{24}Cr$	γ, X-rays
cobalt-60	$^{60}_{27}Co$	β, γ
iodine-131	$^{131}_{53}I$	β
radon-222	$^{222}_{86}Rn$	α
uranium-238	$^{238}_{92}U$	α, β, γ

Critical Thinking Questions

11. Write the symbol from Table 1 (showing the superscript and subscript) for the radioactive particle released by iodine-131.

12. For iodine-131, what is the atomic number? _____ the mass number? _____

13. Consider the incomplete nuclear equation for the decay of iodine-131 shown below. Into which of the two boxes would the particle in CTQ 11 be placed: **left** or **right** (circle one)

$$^{131}_{53}I \; + \; \boxed{} \; \rightarrow \; \boxed{} \; + \; new\ element$$

   a. Draw an "X" through the incorrect box. Explain your team's choice in a sentence.

   b. Predict the mass number of the new element. _____

   c. Predict the atomic number of the new element. _____

   d. Explain why the answer to (c) above is not 52.

   e. Using the information above and a periodic table, what is the symbol of the new element formed?

   f. Write a complete nuclear reaction for the radioactive decay of iodine-131.

Model 3: Amount of iodine-131 remaining as it decays over a two-month period

The time required for half of a sample of a radioactive isotope to decay is called
the **half-life (t½)**.

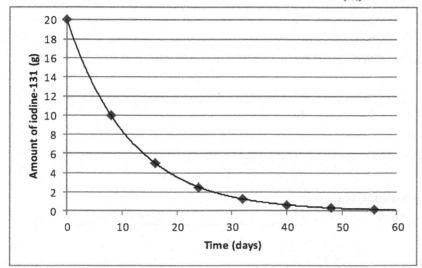

Amount of ^{131}I (g)	Time (d)	Number of half lives
20	0	0
10	8	1
5	16	2
2.5	24	3
1.25	32	4
0.625	40	5
0.313	48	6
0.156	56	7

Critical Thinking Questions:

14. What *fraction* of a sample of iodine-131 is left after 8 days? _____ After 16 days?_____

15. Consider the 20-gram sample of iodine-131 for which data is shown in Model 2.
    a. How many half-lives will it take for the sample to decay until **less than 1%** of the original isotope remains?

    b. How many days would this be?

16. As a team, write a consensus explanation about what happens to the iodine-131 that is "lost" (recall CTQ 13f).

17. As a team, predict when **all** (100%) of the radioactivity from iodine-131 will be lost.

18. Consider a 100-gram sample of radioactive cobalt-60 with a half-life of 5.3 years. Approximately how many grams of radioactive cobalt-60 will remain after 11 years?

19. Considering <u>only the half lives</u> of uranium-238 (4.47 × 10⁹ years) and iodine-131, which would be more appropriate for internal usage (ingestion) for medical tests? As a team, write a consensus explanation for your choice.

20. Explain how your team ensured that each member understood the material as you worked through the activity today.

21. List one insight that your team discovered today about nuclear chemistry.

Information:

Table 3: Half-lives of some radioisotopes

Radioisotope	Radiation type	Half-life	Use
barium-131	γ	11.6 days	detection of bone tumors
carbon-14	β	5730 yr	carbon dating
chromium-51	γ, X-rays	27.8 days	measuring blood volume
cobalt-60	β, γ	5.3 yr	food irradiation, cancer therapy
iodine-131	β	8.1 days	hyperthyroid treatment
uranium-238	α, β, γ	4.47×10^9 yr	dating igneous rocks

Exercises:

1. What kind of personal protective equipment would be appropriate for a person working with a sample of iodine-131 to wear?

2. Predict the new element formed when carbon-14 emits a β particle.

3. Write the nuclear reaction for the radioactive decay of carbon-14.

4. After an organism dies, it stops taking in radioactive carbon-14 from the environment. If the carbon-14:carbon-12 ratio ($^{14}_{6}C / ^{12}_{6}C$) in a piece of petrified wood is one sixteenth of the ratio in living matter, how old is the rock? (Hint: How many half lives have elapsed?)

5. Would chromium-51 be useful for dating rocks containing chromium? Why or why not?

6. Suppose that 0.50 grams of barium-131 are administered orally to a patient. Approximately how many <u>milligrams</u> of the barium would still be radioactive two months later?

7. Complete the equations.

    a. $^{30}_{15}P \longrightarrow ^{0}_{1}e +$ (What type of radiation is this?)

    b. $^{113}_{47}Ag \xrightarrow{\text{beta decay}}$

    c. $\xrightarrow{\alpha \text{ and } \gamma \text{ emission}} ^{222}_{86}Rn + + ^{0}_{0}\gamma$

8. Read the assigned pages, and work the assigned problems.

Electron Arrangement*
(Where are the electrons?)

Information: Ionization Energy

Electrons are attracted to the nucleus due its opposite charge. Therefore, energy must be supplied to pull an electron away from the nucleus, leaving a positively charged species (cation) and a free electron. The **ionization energy** (IE) is the minimum energy required to remove an electron from an atom. Ionization energies are usually obtained experimentally.

Critical Thinking Questions:

1. How many electrons does lithium (Li) have?

2. If electrons are distributed around the nucleus at various distances:
   a. Is the ionization energy for all of them the same?

   b. If not, which electron would have the lowest ionization energy: the electron closest to the nucleus or the electron that is farthest from nucleus?

3. a. As the atomic number increases, does the charge in the nucleus get stronger or weaker?

   b. Would it require more or less energy to remove an electron as the charge changes?

4. Work as a team to predict the relationship between IE and atomic number by making a rough graph of IE vs. atomic number. Do not proceed until you have finished your graph.

IE

Atomic Number

---

* Adapted from ChemActivity 4, Moog, R.S.; Farrell, J.J. *Chemistry: A Guided Inquiry*, 5th ed., Wiley, 2011, pp. 22-28.

Information

One would expect that the ionization energy of an atom would increase as the nuclear charge increases. Also, the IE of an atom should decrease if the electron being removed is further from the nucleus. Table 1 presents the experimentally measured IE's of the first 20 elements.

Table 1. First ionization energies of the first 20 elements.

Atomic Number	Symbol	IE (MJ/mol)	Atomic Number	Symbol	IE (MJ/mol)
1	H	1.31	11	Na	0.50
2	He	2.37	12	Mg	0.74
3	Li	0.52	13	Al	0.58
4	Be	0.90	14	Si	0.79
5	B	0.80	15	P	1.01
6	C	1.09	16	S	1.00
7	N	1.40	17	Cl	1.25
8	O	1.31	18	Ar	1.52
9	F	1.68	19	K	0.42
10	Ne	2.08	20	Ca	0.59

Model 1: The Shell Model for Electrons

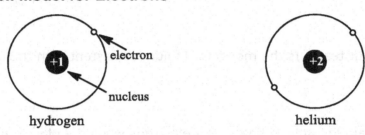

hydrogen   helium

The hydrogen electron "sees" a +1 nuclear charge (from the proton), while each helium electron "sees" a +2 charge. The IE of hydrogen is 1.31 MJ/mole, and the IE of helium is almost twice the amount, consistent with each of the two electrons orbiting the nucleus at a distance approximately the same as that in H. These electrons are in a **shell** around the nucleus.

Critical Thinking Questions:

5. Based on the IE for hydrogen and helium, what would you predict the IE would be for lithium with three electrons? What is the actual value?

6. As a team, predict an arrangement of electrons for lithium that would be consistent with the data.

Model 2: The Shell Model for Lithium

The ionization energy for Li is less than that of He. Actually, it is significantly smaller than that of H. This data is not consistent with placing the third electron in the first shell, or the ionization energy would have been higher than He. Therefore, the electron being removed must be farther from the nucleus, or in a *second shell*.

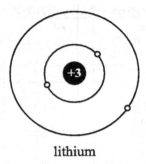

lithium

Critical Thinking Questions:

7. How many total electrons does Li have?

8. Identify the second shell in Model 2. How many electrons were placed in the second shell?

9. Consult with your team. Is the model for lithium consistent with the data? Why or why not?

10. Based on the data in Table 1, how many electrons would be placed in the second shell for lithium?

11. Based on the data in Table 1, with which atom do electrons start entering a third shell? With which atom does a fourth shell start? Does your team agree?

12. Considering your team's answer to CTQ 11, do you observe any trends between electrons being placed in higher shell numbers and the location of the atoms in the Periodic Table?

13. Discuss with your team the overall trends in ionization energy across the Periodic Table and down the Periodic Table.

14. Review the activity, and describe concisely the major concept(s) of this activity.

15. List one area in which your team could use improvement and how you will accomplish that during the next class period.

Exercises:

1. How many electrons are in shell 2 of the following elements?
   a. sodium   b. helium   c. nitrogen

2. How many electrons are in shell 3 of the following elements?
   a. oxygen   b. sulfur   c. phosphorus

3. Indicate how many electrons are in each shell for the following atoms.
   Example: sodium: 2, 8, 1
   a. carbon   b. argon   c. silicon   d. nitrogen

4. Which atom would have a higher ionization energy: K or As? Explain your reasoning.

5. Which atom would have a lower ionization energy: P or Sb? Explain your reasoning.

6. Read the assigned pages in your text, and work the assigned problems.

Valence Electrons
(Where are the electrons in atoms located?)

Model 1: Diagrams of atoms of three elements using the shell model

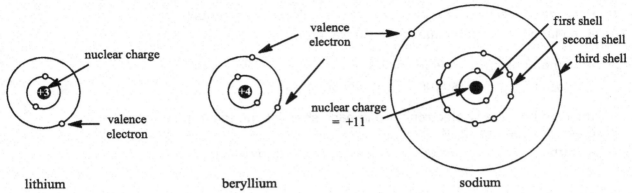

lithium   beryllium   sodium

The **core** of an atom is everything **except** the outermost shell of electrons.

Critical Thinking Questions:

1. How many electrons are in the outermost shell of Li? _____ of Be? _____ of Na? _____

2. How many inner shell (core) electrons does Li have? _____ Be? _____ Na? _____

3. How many protons are in the core of the lithium atom? ____ How many electrons are in the core? ____ What is the net (total) charge in the core?

4. Discuss with your team, and write a sentence to explain how you arrived at your answer to CTQ 3.

5. What is indicated by the term "valence electron"?

Model 2: Incomplete diagrams of a magnesium (Mg) atom using the shell model (a) and the core charge concept (b)

When main group atoms combine to form compounds, they gain, lose, or share electrons in their valence shells. Atoms in stable compounds normally have full valence shells (eight electrons) in all the atoms in the compound. This is commonly known as the **octet rule**.

(a)   (b)

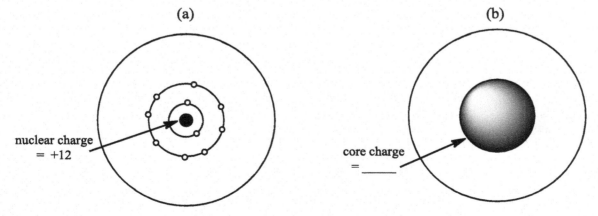

6. Label the shells in Model 2(a) as 1, 2, 3, etc.

7. Why is the nuclear charge of Mg "+12"?

8. Considering Models 1 and 2, how many electrons completely fill the first shell? _____

9. How many electrons fill the second shell? _____

10. How many inner shell (core) electrons does Mg have? _____

11. How many total electrons should Mg have, based upon its atomic number? _____

12. If Mg has the number of electrons you indicated in CTQ 11, how many valence electrons will Mg have? _____ Add them onto the shell model (a) in the diagram above.

13. Work with your team to complete diagram (b) of a magnesium (Mg) atom using the core charge concept.

14. Look back at your answer to CTQ 5. Discuss with your team and refine the meaning of the term **valence electron**. When your team agrees, write a definition in one complete sentence.

15. Review Model 1 and CTQ 14. What do H, Li and Na (and all group 1 elements) have in common?

16. What do Mg and all group 2 elements have in common?

17. As a team, predict whether Mg will likely gain, lose, or share electron(s) to achieve a full octet when it forms a compound. How many electron(s) will be gained or lost?

18. What was your plan for improving performance today? Explain why was your plan was or was not successful.

Exercises:

1. Why are there only two elements in the first row of the periodic table?

2. How many valence electrons do the following atoms contain: Cl, Rb, P, Ba, O?

3. For the following atoms, predict the number of electrons that would be gained or lost to form an octet in the outermost shell: K, N, Br, Al.

4. Read the assigned pages in your text, and work the assigned problems.

Electron Configuration and the Periodic Table
(How are atoms and elements classified?)

Model 1: Electron Arrangement

A person's postal address contains broad categories (state, city) and also gets quite specific (street, house or apartment number). Similarly, electrons in atoms are assigned an "address." The shell number gives a general indication of the location of an electron around a nucleus. However, to accurately describe the locations of each of the electrons, a more precise system is needed. The shell is composed of up to four **subshells** (labeled **s, p, d, f**), and each subshell is subdivided further into **orbitals** holding up to two electrons each. Electrons are arranged in addresses beginning at the lowest level closest to the nucleus and continuing until the total number of electrons is reached.

Table 1. Electron capacity for subshells.

shell	subshell	# of orbitals	orbital names
1	s	1	1s
2	s	1	2s
	p	3	2p 2p 2p
3	s	1	3s
	p	3	3p 3p 3p
	d	5	3d 3d 3d 3d 3d
4	s	1	4s
	p	3	4p 4p 4p
	d	5	4d 4d 4d 4d 4d
	f	7	4f 4f 4f 4f 4f 4f 4f

Critical Thinking Questions:

1. How many subshells does shell number 1 have? ____ How many total orbitals? ____

2. How many subshells does shell number 2 have? ____ How many total orbitals? ____

3. How many subshells does shell number 3 have? ____ How many total orbitals? ____

4. Each orbital can hold two electrons. How many total electrons can the first shell hold?___

5. How many electrons can the second shell hold? ____

6. a. How many electrons does a carbon (C) atom have? ____

   b. How many orbitals are required to hold all the electrons in a carbon atom? ____

   c. Starting from the first shell and proceeding in order that orbitals are listed in Table 1, work together as a team to determine which orbitals would hold the electrons in a carbon atom. Write the orbital names in order.

7. a. How many electrons does a nitrogen (N) atom have? ____

   b. How many orbitals will N need to hold its electrons? ____

   c. Write the orbital names (in order) that would be used to hold the electrons in N.

Model 2: Electron configurations for selected elements

Element	Configuration	Element	Configuration
H	$1s^1$	Na	$1s^2\ 2s^2\ 2p^6\ 3s^1$
He	$1s^2$	Mg	$1s^2\ 2s^2\ 2p^6\ 3s^2$
Li	$1s^2\ 2s^1$	Al	
Be		Si	$1s^2\ 2s^2\ 2p^6\ 3s^2\ 3p^2$
B	$1s^2\ 2s^2\ 2p^1$	P	$1s^2\ 2s^2\ 2p^6\ 3s^2\ 3p^3$
C	$1s^2\ 2s^2\ 2p^2$	S	$1s^2\ 2s^2\ 2p^6\ 3s^2\ 3p^4$
N	$1s^2\ 2s^2\ 2p^3$	Cl	$1s^2\ 2s^2\ 2p^6\ 3s^2\ 3p^5$
O		Ar	$1s^2\ 2s^2\ 2p^6\ 3s^2\ 3p^6$
F		K	
Ne	$1s^2\ 2s^2\ 2p^6$		

Critical Thinking Questions:

8. Compare your answers to CTQs 6 and 7 with the electron configurations for C and N in Model 2. What do the superscript numbers indicate in the notation in the Table?

9. As a shorthand notation, electrons in subshells of the same type are written together. For example, five electrons in three 2p subshells ($2p^2\ 2p^2\ 2p^1$) is written "$2p^5$." How would 6 electrons in three 3d subshells be written? _____

10. What is the major difference between the electron configurations of Ne and Na?

11. How many electron shells does Ne have? ____ Na? ____

12. Based on the electron configuration of an atom, how can you determine how many electron shells an atom has? Discuss this with your team, and write your answer in a complete sentence.

13. Work together as a team to complete the missing electron configurations in Model 2.

Model 3: Arrangement of the periodic table

Notice that the Periodic Table is arranged in a specific block format with two columns on the left, ten in the middle, and six on the right. The inner transition elements actually belong in rows 6 and 7, so there are seven rows in the Periodic Table.

group:	1	2	3	4	5	6	7	8	9	10	11	12	13	14	15	16	17	18
shell 1	1 H						Main group elements (excludes groups 3-12)											2 He
2	3 Li	4 Be											5 B	6 C	7 N	8 O	9 F	10 Ne
3	11 Na	12 Mg				Transition Elements							13 Al	14 Si	15 P	16 S	17 Cl	18 Ar
4	19 K	20 Ca	21 Sc	22 Ti	23 V	24 Cr	25 Mn	26 Fe	27 Co	28 Ni	29 Cu	30 Zn	31 Ga	32 Ge	33 As	34 Se	35 Br	36 Kr
5	37 Rb	38 Sr	39 Y	40 Zr	41 Nb	42 Mo	43 Tc	44 Ru	45 Rh	46 Pd	47 Ag	48 Cd	49 In	50 Sn	51 Sb	52 Te	53 I	54 Xe
6	55 Cs	56 Ba	57 La	72 Hf	73 Ta	74 W	75 Re	76 Os	77 Ir	78 Pt	79 Au	80 Hg	81 Tl	82 Pb	83 Bi	84 Po	85 At	86 Rn
7	87 Fr	88 Ra	89 Ac	104 Rf	105 Db	106 Sg	107 Bh	108 Hs	109 Mt	110 Ds	111 Rg	112 Cn	113 Nh	114 Fl	115 Mc	116 Lv	117 Ts	118 Og

Inner Transition Elements

58 Ce	59 Pr	60 Nd	61 Pm	62 Sm	63 Eu	64 Gd	65 Tb	66 Dy	67 Ho	68 Er	69 Tm	70 Yb	71 Lu
90 Th	91 Pa	92 U	93 Np	94 Pu	95 Am	96 Cm	97 Bk	98 Cf	99 Es	100 Fm	101 Md	102 No	103 Lr

Critical Thinking Questions:

14. What correlation exists between the number of electron shells an atom has (see CTQ 12) and the row number in which the element is found in the Periodic Table?

15. How many electron shells does potassium (K) have? ____ Radium (Ra)? ____

16. How many **electrons** does Mg have? ____ Ar? ____

17. Ar has six more electrons than Mg. What is the major difference in the electron configurations of Mg and Ar? (Refer to Model 2).

18. a. How many electrons can the 2s orbital hold? _____ (Refer to Model 1 if necessary.)

    b. How many electrons can all 3 of the 2p orbitals hold in total? _____

19. The Periodic Table in Model 3 is divided into "blocks." Consider the second row, which spans two blocks and contains the elements Mg and Ar. What correlation exists between the number of columns in a block (outlined in bold) and the number of electrons that a particular subshell can hold?

20. Label the "s-block" and the "p-block" in Model 3.

21. Refer to Model 1. How many electrons is the entire 3d subshell (all 5 3d orbitals) capable of holding? ____ The entire 4f subshell? ____ Use this information to label the "d-block" and the "f-block" in Model 3.

22. Work with your team to compare the electron configurations of elements in the same column of the Periodic Table (e.g., H, Li, Na, and K). What do they have in common?

Model 4: Electron configuration of bromine

After calcium (atomic number 20), electrons do not fill neatly in order from shell $1 \rightarrow 2 \rightarrow 3$ *etc.* Consider, for example, the order that orbitals are filled to reach the electron configuration of the element bromine (Br), with 36 total electrons:

$$1s^2\, 2s^2\, 2p^6\, 3s^2\, 3p^6\, 4s^2\, 3d^{10}\, 4p^5$$

After the 4s orbital, the next electrons go into the 3d orbital, which is in the core.

Critical Thinking Questions:

23. Considering the electron configuration of bromine, the 21st through 30th electrons went into the 3d subshell. Why would these electrons not be considered valence electrons?

24. Find iron (Fe) on the periodic table in Model 3.
    a. In what period (row) is it located? ____
    b. How many electrons does it have? ____

25. Following the example above, predict the electron configuration for Fe.

26. How many valence electrons does Fe have? ____

27. Work with your team to explain why the answer to question 21 is **not** 6.

28. Explain why all transition elements should have the same number of valence electrons.

29. Work as a team to identify the major concepts in this activity.

30. Identify one unanswered question that your team has about electron configuration.

Exercises:

1. What would be the maximum number of electrons in the following?
   a. a 2p orbital ____   b. the 2p subshell ____   c. shell 2 ____   d. a 3s orbital

2. How many electrons are present in an atom with its 1s, 2s, and 2p subshells filled? ____ What is the element? ____

3. Write the electron configurations for Ca and Ga.

4. A student makes the following statement: "The electron configuration of all halogens ends in p^5." Is the student correct? Explain.

5. Learn to use the periodic table to locate the main block elements, transition elements, and inner transition elements (see Model 3).

6. Learn the family names of the elements in groups 1, 2, 17 and 18.

7. Read the assigned pages in your text, and work the assigned problems.

Ions and Ionic Compounds
(*How are ionic compounds formed?*)

Model 1: Common charges (by group) on elements *when in ionic compounds*.

Charge

	←				variable charge					→				$\otimes$			$\otimes$

Grp #

1	2	3	4	5	6	7	8	9	10	11	12	13	14	15	16	17	18
1 **H**																	2 **He**
3 **Li**	4 **Be**											5 **B**	6 **C**	7 **N**	8 **O**	9 **F**	10 **Ne**
11 **Na**	12 **Mg**											13 **Al**	14 **Si**	15 **P**	16 **S**	17 **Cl**	18 **Ar**
19 **K**	20 **Ca**	21 **Sc**	22 **Ti**	23 **V**	24 **Cr**	25 **Mn**	26 **Fe**	27 **Co**	28 **Ni**	29 **Cu**	30 **Zn**	31 **Ga**	32 **Ge**	33 **As**	34 **Se**	35 **Br**	36 **Kr**
37 **Rb**	38 **Sr**	39 **Y**	40 **Zr**	41 **Nb**	42 **Mo**	43 **Tc**	44 **Ru**	45 **Rh**	46 **Pd**	47 **Ag**	48 **Cd**	49 **In**	50 **Sn**	51 **Sb**	52 **Te**	53 **I**	54 **Xe**
55 **Cs**	56 **Ba**	57 **La**	72 **Hf**	73 **Ta**	74 **W**	75 **Re**	76 **Os**	77 **Ir**	78 **Pt**	79 **Au**	80 **Hg**	81 **Tl**	82 **Pb**	83 **Bi**	84 **Po**	85 **At**	86 **Rn**
87 **Fr**	88 **Ra**	89 **Ac**	104 **Rf**	105 **Db**	106 **Sg**	107 **Bh**	108 **Hs**	109 **Mt**	110 **Ds**	111 **Rg**	112 **Cn**	113 **Nh**	114 **Fl**	115 **Mc**	116 **Lv**	117 **Ts**	118 **Og**

A <u>cation</u> has a positive charge.   An <u>anion</u> has a negative charge.

Critical Thinking Questions:

1. Recall the shell model of the atom.

   a. How many electrons fill the first shell? _____

   b. How many electrons fill the second shell? _____

2. How many valence electrons are found in atoms in

   a. group 1? _____   b. group 13? _____   c. group 16? _____

3. When atoms form ionic compounds, they often fill their outer valence shell (**octet rule**) by either gaining or losing 1, 2 or 3 electrons. Based on your answer to CTQ 2, indicate how many electrons these atoms would either gain or lose to achieve an octet.

   a. group 1? _____   b. group 13? _____   c. group 16? _____

4. What would be the resulting charge on the ions created in CTQ 3 when electrons are gained or lost?

   a. group 1? _____   b. group 13? _____   c. group 16? _____

5. Based on your answer to CTQ 4, do metals typically form **anions** or **cations** (circle one)?

6. Work with your team and write charges in the empty boxes above each group (skip the transition metals, groups 3-12) at the top of the Periodic Table in Model 1.

7. Do nonmetals typically form **anions** or **cations** (circle one)?

Model 2: Compound formation

Atoms can form ions when donating or accepting valence electrons to or from another atom. These ions are held together by strong electrical interactions between opposite charges, called an **ionic bond**, and thus forming a neutral **compound**.

Figure 1. Valence electron donation from lithium to bromine.

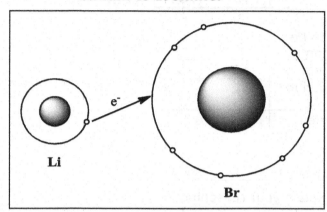

Figure 2. Lithium bromide compound.

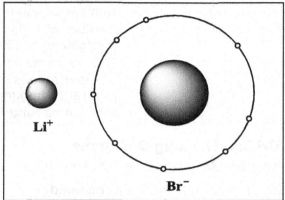

Critical Thinking Questions:

8. How many electrons did lithium donate? _____ How many did bromine accept? _____

9. Based on the electron configuration, which noble gas does lithium now resemble? _____

10. Based on the electron configuration, which noble gas does bromide now resemble?_____

11. Both the lithium ion and the bromide ion have charges. What is the **net** (total) charge on the compound (LiBr) that is formed in Figure 2? _____ Ensure that all team members understand how you arrived at this answer.

12. a. An oxygen atom can **lose** or **gain** (circle one) two electrons to achieve a noble gas configuration.

    b. How many electrons will this oxygen ion have? _____ How many protons? _____

    c. What will be the charge on this oxygen ion? _____

    d. Following the examples of Li^+ and Br^- in Figure 2, write the symbol for the ion formed from an oxygen atom. _____

    e. Based on the electron configuration, which noble gas does an oxygen ion resemble?

13. Lithium can also form compounds with oxygen. Work with your team and determine which of the three formulas shown below would represent the neutral compound that would be formed. Circle your choice, and explain why you chose it.

$$LiO \qquad\qquad Li_2O \qquad\qquad LiO_2$$

Information

Ionic compounds are always formed by the interaction of cations (typically metals) with anions (nonmetals). The number of each ion in the compound is adjusted so that the resulting compound has an overall charge of zero.

Table 1. Steps in determining the likely formula of aluminum oxide.

	Al	O
Number of valence e⁻	3	6
Number of e⁻ likely to be lost/gained to form ion	3	2
Charge on resulting ion	+3	−2
Number of ions needed for neutral compound	2	3
Ionic compound formed	Al_2O_3	

Critical Thinking Questions

14. Consider Table 1. Will Al lose or gain electrons? _____ O? _____

15. Is Al_2O_3 is a neutral compound? _____ Explain your reasoning.

16. Complete the following table, beginning with the top row and proceeding to each row in sequence.

	Ca	F
Number of valence e⁻		
Number of e⁻ likely to be lost/gained to form ion		
Charge on resulting ion		
Number of ions needed for neutral compound		
Ionic compound formed		

17. Iron (Fe) is a transition metal and can form two possible ions. If iron reacts with oxygen and forms FeO, what is the charge on the iron? _____ What is the charge on iron in Fe_2O_3? _____ Does your team agree?

18. What did the manager do today to ensure that everyone understood the material? How did he or she make sure that everyone worked together?

Exercises:

1. Based on their electron shell configurations, give an explanation for why all alkaline earth metals in ionic compounds have a +2 charge.

2. Based on their electron shell configurations, give an explanation for why all halogens in ionic compounds have a −1 charge.

3. Indicate whether the following statements are true or false and explain your reasoning.
    a. Magnesium donates two electrons when it forms an ion.

    b. Iodine forms an ion with a −2 charge.

    c. The correct formula for the ionic compound formed between potassium and chlorine is KCl_2.

4. What ions are formed by sodium and sulfur? What compound would be formed by these ions?

5. Circle each of the following pairs of elements that are likely to form an ionic compound.

    a. lithium and chlorine   c. magnesium and neon

    b. sodium and calcium   d. potassium and oxygen

6. Write the correct formula for ionic compounds formed between the following pairs of elements.
    a. aluminum and bromine

    b. magnesium and nitrogen

7. Read the assigned pages in the text, and work the assigned problems.

Naming Ionic Compounds
(*How are ionic compounds named*?)

Model 1: Formulas and names of some binary ionic compounds

Formula	Group Number of Metal	Name
NaBr	1	sodium bromide
K_2O	1	potassium oxide
MgF_2	2	magnesium fluoride
$CrCl_4$	6	chromium(IV) chloride
FeS	8	iron(II) sulfide
Fe_2S_3	8	iron(III) sulfide
Cu_2S	11	copper(I) sulfide
InP	13	indium phosphide
PbI_2	14	lead(II) iodide

In $CrCl_4$, the cation is named <u>chromium(IV)</u> and the anion is named <u>chloride</u>.

Critical Thinking Questions:

1. In the name of an ionic compound, which ion is always written first: **the anion** or **the cation** (circle one).

2. Look at the names of the compounds in Model 1. What difference, if any, is there between the name of the *anion* and the name of the nonmetallic element that it corresponds to? Write your answer in a complete sentence.

3. In Model 1, circle the **names** of compounds for which the metal ion is **not** in groups 1, 2, or 13. Ensure that all team members have the same names circled.

4. Look at the names you circled in CTQ 3. What difference, if any, is there between the name of the *cation* and the name of the metallic element that it corresponds to? Discuss with your team, and when you agree, write an explanation in a complete sentence.

5. Look at the names you did **not** circle in CTQ 3. For metal ions from groups 1, 2, or 13 in ionic compounds, how is the name of the cation related to the name of the metal? Discuss with your team and explain in a sentence.

6. Consider the formula for magnesium fluoride, MgF_2.

   a. What is the charge on the magnesium ion? _____

   b. What is the charge on *each* fluoride ion? _____

   c. What is the overall (total) charge on the compound? _____

7. Find three other compounds in Model 1 that are not circled. Write the formula of the compound and follow the process in CTQ 6 to calculate the overall charge on the compound.

      Formula: _____   Charge: _____

      Formula: _____   Charge: _____

      Formula: _____   Charge: _____

8. In general, what is the overall charge on an ionic compound? _____

9. Based on your answer to CTQ 8, what must the charge be on each iron atom in FeS?_____ In Fe_2S_3? _____

10. Based on your answer to CTQ 9, what do the Roman numerals in Model 1 represent? Ensure that all team members agree.

11. As a team, fill in the blanks to complete the rules for naming ionic compounds:

   - Naming metal ions: name the metal [example: Ca^{2+} = _____]

   - If the metal is **not** in group 1, 2, or 13, add a Roman numeral in parentheses that

     represents _____[e. g., Fe^{3+} =_____]

   - Nonmetals: change ending of element name to _____[e. g., N^{3-} = _____]

   - Naming ionic compounds: name the cation, then the anion [example: FeN =

     _____]

12. List any strategies for naming compounds that your team discovered today.

13. List one strength of your team work today and why it helped your understanding.

Information

Table 1: Formulas and names of some common polyatomic ions to memorize

Formula	Name	Formula	Name
NH_4^+	ammonium	NO_3^-	nitrate
H_3O^+	hydronium	SO_4^{-2}	sulfate
$C_2H_3O_2^-$	acetate	PO_4^{-3}	phosphate
CN^-	cyanide	MnO_4^-	permanganate
OH^-	hydroxide	CO_3^{-2}	carbonate
ClO_3^-	chlorate	$Cr_2O_7^{-2}$	dichromate

Table 2: Rules for naming other polyatomic ions

Rule	Examples
1. If adding a H^+ to a polyatomic ion results in a new ion, add the word "hydrogen" in front of the name; if adding 2 H^+, add the word "dihydrogen."	HPO_4^{2-} = hydrogen phosphate $H_2PO_4^-$ = dihydrogen phosphate
2. For ions with one fewer oxygen atoms than the common ion, change the ending from "-ate" to "-ite"	SO_4^{2-} = sulf**ate**, so SO_3^{2-} = sulf**ite**

Exercises:

1. Complete the following table.

Formula	Name
$CoCl_4$	_____
_____	potassium nitrate
$Ba(OH)_2$	_____
_____	sodium hydrogen carbonate
_____	beryllium bromide
Li_2CO_3	_____
_____	copper(II) oxide
_____	sodium hypochlorite
$Ca(C_2H_3O_2)_2$	_____
_____	magnesium sulfate
$NaNO_2$	_____
_____	vanadium(III) sulfate

2. Read the assigned pages in the text, and work the assigned problems.

Covalent Bonds[*]
(Why do atoms share electrons?)

Model 1: Two types of chemical bonding[†]

Ions held together by opposite charges are said to be <u>ionically</u> bonded.
Ionic <u>compounds</u> contain ions—typically a metal ion along with nonmetals.

Atoms <u>sharing</u> valence electrons are said to be <u>covalently</u> bonded.
Covalent <u>molecules</u> typically contain <u>only nonmetals</u>.

Critical Thinking Question:

1. Predict whether a bond between the two given elements would be ionic or covalent.

    a. sulfur and sodium _____

    b. sulfur and hydrogen _____

Model 2: Lewis electron-dot structures for hydrogen and second row elements

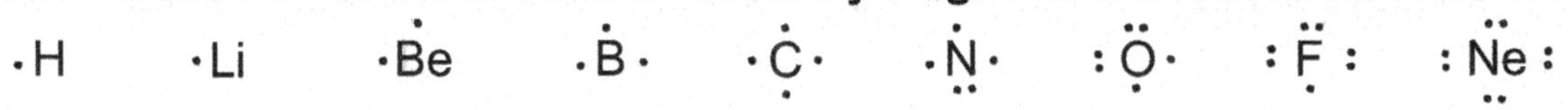

·H   ·Li   ·Be   ·B·   ·C·   ·N·   :O·   :F: :Ne:

Critical Thinking Questions:

2. Which two elements in Model 2 are metals?

3. Are these two metallic elements likely to be in a covalent molecule? Explain your answer. Check that all team members agree on the explanation.

4. Consider Model 2. How is the number of dots related to the number of valence electrons?

5. By extension, write the electron-dot (Lewis) structures for sulfur, chlorine, and sodium atoms.

6. Write electron dot (Lewis) structures for the **ions** S^{2-} and Cl^-. Does your team agree?

---

[*]Adapted from ChemActivity 13, Moog, R.S.; Farrell, J.J. *Chemistry: A Guided Inquiry*, 5th ed., Wiley, 2011, pp. 74-80.
[†] A third type of bonding, metallic bonding, will not be considered in this book.

Critical Thinking Questions:

7. a. How many **electrons** in the dot structure for O in Model 2 are **in pairs**? _____

    b. How many electrons are **unpaired**? _____
    Manager: Have a team member check with another team to see if you agree on answers to parts a and b.

8. How many electrons in the dot structure for N are in pairs? _____ unpaired? _____

Model 3: Typical number of covalent bonds for elements common in biological molecules

Element	Number of Bonds
H	1
O	2
N	3
C	4

Critical Thinking Questions:

9. Considering your answers to CTQs 7-8, how is the number of covalent bonds that an atom makes related to its electron-dot structure?

10. Which nonmetal in Model 2 is *unlikely* to be in a covalent molecule? Discuss with your team and write a sentence to explain.

Model 4: Covalent bonding (sharing valence electrons) between H and F

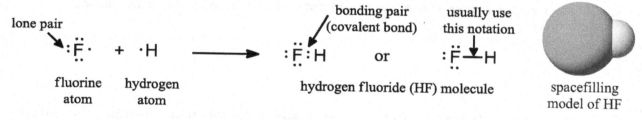

Another example: sharing between H and O

Critical Thinking Questions:
*Refer to the **HF molecule** in Model 4 to answer CTQs 11-14.*

11. In the HF molecule, how many <u>total electrons</u> are being shared by F in bonding pairs? ___

12. How many <u>electrons</u> are in lone pairs on F? ____

13. What is the <u>total</u> number of electron dots in lone pairs around <u>or</u> being shared by F? ____

14. What is the total number of electrons in lone pairs and bonding pairs being shared by H in HF? _____

15. Given the shell model of the atom, why do you think that Lewis proposed a maximum of only **two** electron-dots for hydrogen, but **eight** for carbon, nitrogen, oxygen, and fluorine atoms? Discuss your answer with your team and record your consensus answer.

16. Answer the following for the phosphorus atom:
    a. How many valence electrons does P have? _____
    b. What is the Lewis representation for P? (Note: it does not matter on which side of the element symbol the paired electrons are drawn).

    c. How many additional electrons does P need when it forms a compound? _____
    d. What is the likely <u>formula</u> for a compound composed of hydrogen atoms and one

        phosphorus atom? _____ Draw the Lewis structure.

17. Answer the following for the sulfur atom:
    a. How many valence electrons does S have? _____
    b. What is the Lewis representation for S?

    c. How many additional electrons does S need when it forms a compound? _____
    d. What is the likely formula for a compound composed of hydrogen atoms and one sulfur atom? _____ Would bonds in this compound be **ionic** or **covalent** (circle one)? Explain.

    e. Draw the Lewis structure.

18. a. What is the likely formula for a compound composed of sodium atoms and one sulfur atom?

    b. Would this compound be **ionic** or **covalent** (circle one)? Explain.

    c. Why would it be inappropriate to draw a Lewis structure (such as in Model 4) for this compound?

19. Without attempting to draw a Lewis structure, add up the total number of valence electrons that would be in each of these molecules:

    a. Cl_2_____   b. N_2 _____   c. H_2CO _____

20. How did you determine the answers to CTQ 19?

Model 5: Lewis structures of some molecules with multiple bonds

Molecular Formula	Lewis Structure
Cl_2	$:\ddot{Cl}\!\!-\!\!\ddot{Cl}:$
N_2	$:N\!\!\equiv\!\!N:$
H_2CO	$H\!\!-\!\!C\!\!-\!\!H$ $\|$ $:\ddot{O}:$

Multiple bonds are represented with more than one straight line between atoms. A **double bond** is formed when two atoms share two pairs of electrons and is stronger than a single bond. A **triple bond** is formed when two atoms share three pairs of electrons and is stronger than a double bond.

Critical Thinking Questions:

21. What is the total number of electrons shown in the Lewis structure in Model 5 for each molecule?

    b. Cl_2_____   b. N_2 _____   c. H_2CO _____

22. Compare your answers to CTQs 19 and 21. How does one determine the total number of electrons that should be used to generate a Lewis structure?

23. For N_2, is the sum of the bonding electrons and the lone pair electrons around each N atom consistent with the Lewis model? In other words, does each N have an *octet*? Write a sentence to explain.

24. For H_2CO:
    a.   Is the sum of the bonding electrons and the lone pair electrons around each C atom consistent with a complete octet?

    b.   Is the sum of the bonding electrons and the lone pair electrons around each O atom consistent with a complete octet? _____

25. Using the information from above, construct a checklist that can be used to determine if a Lewis structure for a molecule is correct. Try to think of at least three criteria. Compare your criteria with your team and try to come to agreement.

26. List any questions that remain with your team about Lewis dot structures.

Exercises:

1. Follow the steps below to draw the correct Lewis structure for CO_2.

   a. What is the total number of valence electrons in CO_2? _____

   b. Draw an initial Lewis structure connecting each oxygen atom with a single bond to a central carbon atom. (This may be called a "skeleton structure.")

   c. How many valence electrons are represented in the structure in (b)? _____

   d. How many still need to be added, based on your calculation in (a)? _____

   e. Add the remaining electrons *in pairs* (b) around each oxygen atom to give them octets. If any electrons still remain, place them around the central atom (carbon).

   f. Is the sum of the bonding electrons and the lone pair electrons around each O atom consistent with a complete octet? _____

   g. Is the sum of the bonding electrons and the lone pair electrons around the C atom consistent with a complete octet? _____

   h. If your answers to (f) and (g) are yes, then you are done. If not, redraw a Lewis structure in which one or more lone pairs of electrons are moved to create a double or triple bond with the central atom.

   i. Review your structure in (h). Is the number of valence electrons represented in (h) the same as calculated in (a)? _____ Does each atom have an octet?_____

2. Revise (as necessary) your checklist from CTQ 25 that can be used to determine if a Lewis structure for a molecule is correct. Write your list below.

3. Use your checklist to determine whether or not the following is a correct structure for SO_2. If it is not, correct it.

   :Ö—S=O:

4. Draw reasonable Lewis (electron-dot) structures for the following molecules:

   a. NH_3 (ammonia)

   b. $CHCl_3$ (chloroform)

   c. $COCl_2$ (phosgene); C is central

   d. HNO_3 (nitric acid); N is central, and the H is attached to one of the oxygen atoms.

5. The skeleton structures for formaldehyde (left) and methanol (right) are shown below:

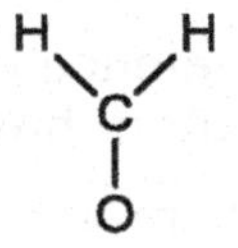

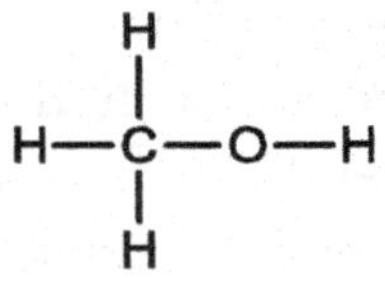

   Consider the following statement:

   It takes more energy to break the C-O bond in formaldehyde than to break the C-O bond in methanol.

   Is the statement **true** or **false** ? (Circle one.) Explain your reasoning. [Hint: Complete the Lewis structures for each molecule.]

6. In the ionic compound NaCl, the sodium has a +1 charge and the chloride has a −1 charge. Based on the Lewis (electron-dot) representations of Na and Cl, hypothesize how these charges could be predicted.

7. Read the assigned pages in the text, and work the assigned problems.

Electrolytes, Acids, and Bases
(Which compounds produce ions when dissolved in water?)

Suggested Demonstration: Electrolytes

Model 1: Electrolytes

> Only separate, charged particles (such as ions) can carry electrical currents.

> *Electrolytes* are substances that can carry an electrical current when dissolved in water.

Critical Thinking Question:

1. What happens to electrolytes when they dissolve in water? Does your team agree?

Model 2: Types of electrolytes

> In water solution, *strong electrolytes* dissociate completely into ions, *weak electrolytes* dissociate only slightly, and *nonelectrolytes* dissociate undetectably or not at all.

Critical Thinking Question:

2. Work with your team to describe a method by which you could tell if a particular solution contains a strong electrolyte, weak electrolyte, or nonelectrolyte. Write the answer in a complete sentence. Record your team answer on the report form, if applicable.

Model 3: Some common acids and bases

Type of electrolyte	Acids	Bases
Strong (any acid or base not listed here is considered weak)	HCl	LiOH
	HBr	NaOH
	HI	KOH
	H_2SO_4	$Ca(OH)_2$
	HNO_3	$Sr(OH)_2$
	$HClO_4$	$Ba(OH)_2$
Weak	$HC_2H_3O_2$	$Mg(OH)_2$
	HCN	

Acids dissociate in water to give hydrogen (H^+) ions and an anion. Bases dissociate in water to give hydroxide (OH^-) ions and a cation. *Strong* acids and bases dissociate *completely*. Note that the common acids are molecules, while the bases are ionic compounds.

Critical Thinking Questions:

Manager: *for CTQs 3–7, identify a different team member to give the first explanation.*

3. Consider Model 3. What do all the molecular formulas of the acids have in common?

4. How can you recognize an acid from the molecular formula?

5. How can you recognize a base from its formula?

6. What happens to strong acids when dissolved in water?

7. What happens to weak acids when dissolved in water? Discuss with your team and prepare to share the team answer with the class.

8. Are all acids strong electrolytes? Discuss with your team and write the team's explanation. Record your team's answer (on the report form, if applicable).

9. Describe what happens to the ions in solid sodium hydroxide (NaOH) during the process of dissolving in water.

10. Water is, in actuality, an **extremely weak** acid (see CTQ 4). Explain, then, how water is often considered to be a nonelectrolyte. (You may wish to refer to Models 1 and 2).

11. Comment on strengths and needed improvements for your teamwork so far.

Model 4: Solubilities of some ionic compounds

Ionic compound	Solubility in water	Type of electrolyte	Major species present when dissolved in water
$MgCl_2$	soluble	strong	$Mg^{2+}(aq)$, $Cl^-(aq)$
MgO	insoluble	very weak	$MgO(s)$
K_2S	soluble	strong	$K^+(aq)$, $S^{2-}(aq)$
CuS	insoluble	very weak	$CuS(s)$
$Ca_3(PO_4)_2$	insoluble	very weak	$Ca_3(PO_4)_2(s)$
$Ca(NO_3)_2$	soluble	strong	$Ca^{2+}(aq)$, $NO_3^-(aq)$

Molecular compounds (other than acids) <u>do not dissociate</u> *significantly in water and so are* <u>nonelectrolytes</u>.

The phase label *(aq)*, meaning "aqueous," can be used to show that a species is dissolved in water. **We will not consider how to <u>predict</u> if an ionic compound is soluble in water until later in the course.**

Critical Thinking Questions:

12. A solution of $MgCl_2$ in water could be written as $MgCl_2(aq)$. Besides water, what chemical species would actually be present in the solution?

13. MgO is an ionic compound that is not soluble in water. This means that it would just collect as a solid—*i. e.*, $MgO(s)$—at the bottom of the container. Would there be **any** ions dissolved in the water? Discuss with your team and when all agree, write an explanation. Record on the report form if you have one.

14. Describe a method by which you could determine if an ionic compound is soluble in water. Be prepared to share your team's answer with the class.

15. Consider Model 4. How is the solubility of an ionic compound related to its classification as an electrolyte?

16. The following compounds from Model 4 are mixed with water. Add the phase labels (s) or (aq) to represent whether the compounds are soluble or not:

    K_2S CuS $Ca(NO_3)_2$ $Ca_3(PO_4)_2$

17. Suppose that each compound in the table below is placed in water. The phase labels given describe whether the compound dissolves or not. Fill in the table with the properties for each compound. ***Manager:*** rotate members of your team to give the first answer.

Compound	Identification (choose one): **Acid**, other **molecule, base,** or other **ionic compound**	**Strong, weak,** or **non**electrolyte
$HBr(aq)$		
$KOH(aq)$		
$CH_2O(aq)$		
$Ca_3(PO_4)_2(s)$		
$Al(OH)_3(s)$		
$HOCl(aq)$		
$H_2S(aq)$		
$Fe_2O_3(s)$		
$Na_2S(aq)$		

18. How did your team members perform their roles today? How might you perform better next time? Record any suggestions for improvement.

Exercises:

1. Write formulas for the major species present in the solutions from CTQ 17.

Compound	Major species present when dissolved in water (or "need more information")
$HBr(aq)$	
$KOH(aq)$	
$CH_2O(aq)$	
$Ca_3(PO_4)_2(s)$	
$Al(OH)_3(s)$	
$HOCl(aq)$	
$H_2S(aq)$	
$Fe_2O_3(s)$	
$Na_2S(aq)$	

2. Read the assigned pages in your text, and work the assigned problems.

Naming Binary Molecules, Acids, and Bases
(How are chemical formulas and names related?)

Information: Naming Binary Molecules

Recall that <u>molecules</u> (as opposed to <u>ionic compounds</u>) are formed from nonmetals only. There is a simple method to name binary molecules (molecules containing only two elements).

1. Name the first element, using a prefix to indicate the number of atoms in the formula.
2. Name the second element, using a prefix to indicate the number of atoms in the formula, and changing the ending to "-ide."
3. Remove the vowel at the end of the prefix if keeping it would result in "ao" or "oo".
4. Do not use the prefix "mono-" on the first element.

Table 1: Prefixes to indicate the number of atoms of an element in a binary molecule

Number	Prefix		Number	Prefix
1	mono-		6	hexa-
2	di-		7	hepta-
3	tri-		8	octa-
4	tetra-		9	nona-
5	penta-		10	deca-

Examples: NI_3 = nitrogen triiodide; N_2O = dinitrogen monoxide (not "mon**oo**xide"); N_2O_5 = dinitrogen pentoxide (not "pent**ao**xide")

Critical Thinking Questions:

1. The following is from "Ban Dihydrogen Monoxide," Coalition to Ban DHMO, 1988. Quoted in http://www.dhmo.org/truth/Dihydrogen-Monoxide.html [accessed 03 Feb 2006]:

   Dihydrogen monoxide is colorless, odorless, tasteless, and kills uncounted thousands of people every year. Most of these deaths are caused by accidental inhalation of DHMO, but the dangers of dihydrogen monoxide do not end there. Prolonged exposure to its solid form causes severe tissue damage. Symptoms of DHMO ingestion can include excessive sweating and urination, and possibly a bloated feeling, nausea, vomiting and body electrolyte imbalance. For those who have become dependent, DHMO withdrawal means certain death.

   Explain how these claims are true or false.

2. The following molecule names are incorrect. State the number of the rule on the previous page that is violated. Then give the correct name.

   a. NO   nitrogen oxide

   b. CO   monocarbon monoxide

   c. CS_2 carbon disulfate

Information: Naming the Binary Acids

When the binary acids hydrogen chloride (HCl), hydrogen bromide (HBr), and hydrogen iodide (HI) are dissolved in water (aq), they are called "hydrochloric acid," "hydrobromic acid," and "hydroiodic acid" respectively.

Critical Thinking Questions:

3. Based on the names of HCl, HBr, and HI, what would HF(aq) be called?

4. Is HF considered a **strong** or **weak** acid (circle one)? Explain how you know this.

Information: Naming Oxoacids

An **oxoacid** is an acid that contains hydrogen, oxygen, and one other element. Oxoacids may be formed by adding H^+ to polyatomic ions. One H^+ is added for each negative charge on the ion (example: the charge on sulfate is -2, so add two H^+ ions to make H_2SO_4).

   A. if the ion has an ending of "-ate," the acid has an ending of "-ic" (example: SO_4^{2-} = "sulf**ate**," so H_2SO_4 = "sulfur**ic** acid").
   - memory device: "**Ic**k," I "**ate**" the acid!
   B. if the ion has an ending of "-ite," the acid has an ending of "-ous" (example: SO_3^{2-} = "sulf**ite**," so H_2SO_3 = "sulfur**ous** acid").

Critical Thinking Questions:

5. Name the following oxoacids. (You may wish to consult the polyatomic ion names in ChemActivity 12, Tables 1 and 2.)

   a. HNO_3 _____

   b. HNO_2 _____

   c. $HC_2H_3O_2$ _____

   d. $HClO_3$ _____

   e. HClO (or HOCl)   _____

6. Critical thinking requires integrating new concepts with your previous knowledge in order to draw a conclusion. Engage in a team discussion for a few minutes. Attempt to list as many examples as you can of when your team utilized critical thinking while working on this activity.

Information: Summary flowchart for naming binary ionic and molecular compounds

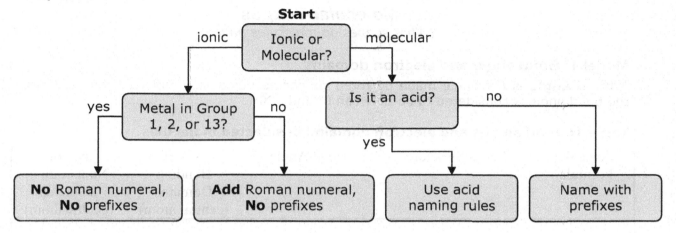

Exercises:

1. Fill in the table with the name of each compound.

Compound	Name
MnS	
CS$_2$	
Na$_2$S	
MgSO$_4$	
H$_2$SO$_3$	
H$_3$PO$_4$	
CuOH	
P$_2$O$_5$	

2. Fill in the table with the formula of each compound.

Compound	Name
	acetic acid
	dinitrogen monoxide
	copper(I) oxide
	carbonic acid
	magnesium bromide
	oxygen difluoride
	sodium nitrite
	tin(II) hydroxide

3. Read the assigned pages in your text, and work the assigned problems.

*Molecular Shapes** *
(What shapes do molecules have?)

Model 1: Bond angle and electron domains.

A **bond angle** is the angle made by three connected nuclei in a molecule. By convention, the bond angle is considered to be between 0° and 180°.

Table 1: Bond angles and electron domains in selected molecules.

Molecular Formula	Lewis Structure	Bond Angle	No. of Bonding Domains (central atom)	No. of Nonbonding Domains (central atom)
CO_2	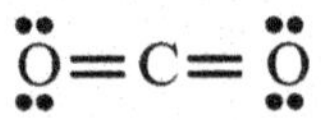 $\ddot{O}=C=\ddot{O}$	∠OCO = 180°	2	0
HCCH	 H—C≡C—H	∠HCC = 180°	2	0
H_2CCCH_2	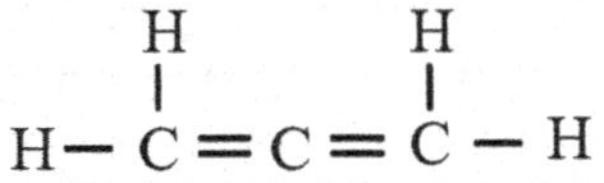 H—C=C=C—H (with H on terminal carbons)	∠CCC = 180°	2	0
ClNNCl	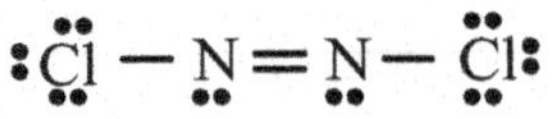 :C̈l—N=N—C̈l:	∠ClNN = 117.4°	2	1
NO_3^-	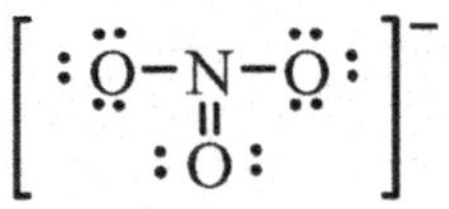 [:Ö—N—Ö: with O below]$^-$	∠ONO = 120°	3	0
H_2CCH_2	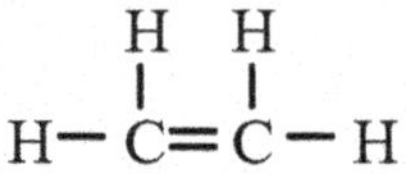 H—C=C—H (with H H on carbons)	∠HCH = 121.1°	3	0

(Table 1 continues on the next page)

---
* Adapted from ChemActivity 18, Moog, R.S.; Farrell, J.J. *Chemistry: A Guided Inquiry*, 5th ed., Wiley 2011, pp. 108-116.

Table 1 (continued): Bond angles and electron domains in selected molecules.

Molecular Formula	Lewis Structure	Bond Angle	No. of Bonding Domains (central atom)	No. of Nonbonding Domains (central atom)
CH_4	H | H — C — H | H	$\angle HCH = 109.45$	4	0
CH_3F	:F̈: | H — C — H | H	$\angle HCH = 109.45°$ $\angle HCF = 109.45°$	4	0
CH_3Cl	:C̈l: | H — C — H | H	$\angle HCH = 109.45°$ $\angle HCCl = 109.45°$	4	0
CCl_4	:C̈l: | :C̈l — C — C̈l: | :C̈l:	$\angle ClCCl = 109.45°$	4	0
NH_3	H — N̈ — H | H	$\angle HNH = 107°$	3	1
NH_2F	H — N̈ — F̈: | H	$\angle HNH = 106.95°$ $\angle HNF = 106.46°$	3	1
H_2O	:Ö — H | H	$\angle HOH = 104.5°$	2	2

Critical Thinking Questions:

1. Fill in the blank in the sentence below with the best term from the following list: **atoms, bonding electrons, lone pairs, nonbonding electrons.**

   The number of bonding domains on a given central atom within a molecule (such as those in Table 1) is equal to the number of _____ attached to the central atom.

2. Fill in the blank in the sentence below with the best term from the following list: **atoms, bonding electrons, lone pairs, nonbonding electrons.**

   The number of nonbonding domains on a given central atom within a molecule (such as those in Table 1) is equal to the number of _____ attached to the central atom.

3. The bond angles in Table 1 can be grouped, roughly, around three values. What are these three values?

4. Based on the numbers of bonding and nonbonding domains in a particular molecule, how can you determine which of the three groupings identified in CTQ 3 it will belong in? Does your team agree?

Model 2: Models for methane, ammonia, and water.

Use a molecular modeling set to make the following molecules: CH_4; NH_3; H_2O. In many model kits: carbon is black; oxygen is red; nitrogen is blue; hydrogen is white; use the short, gray links for single bonds. Nonbonding electrons are not represented in these models.

Critical Thinking Questions:

5. Sketch a picture of the following molecules based on your models: CH_4; NH_3; H_2O.

6. Describe (with a word or short phrase) the shape of each of these molecules: CH_4; NH_3; H_2O. Compare with your team.

7. Consider the water molecule in Table 1. How many bonding domains are around the central oxygen?___ How many nonbonding domains?___ How many total domains?___

Model 3: Types of electron domain geometries.

A domain of electrons (two electrons in a **nonbonding domain**, sometimes called a **lone pair**; two electrons in a **single bond domain**; four electrons in a **double bond domain**; six electrons in a **triple bond domain**) tends to repel other domains of electrons. Domains of electrons around a central atom will orient themselves to minimize the electron-electron repulsion between the domains.

Figure 1. Electron geometry for two, three, and four domains of electrons.

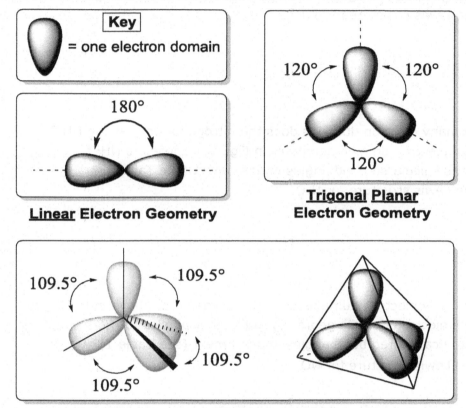

Critical Thinking Questions:

8. Based on Figure 1, what bond angle is expected for a molecule containing:

    a. two domains of electrons? _____

    b. three domains of electrons? _____

    c. four domains of electrons? _____

Manager: *Ensure your team discusses parts a, b, and c for the CTQs 9-13.*

9. Draw the Lewis structure for CH_4 (you may copy it from Table 1).

     a. In total, how many domains of electrons does the carbon atom have in CH_4? _____

     b. Which electron domain geometry in Figure 1 applies to CH_4? _____

     c. Is the geometry in agreement with the calculated bond angles (see Table 1)? _____

10. Draw the Lewis structure for NH_3.

     a. How many electron domains does the nitrogen atom have in NH_3? _____

     b. Which electron domain geometry in Figure 1 applies to NH_3? _____

     c. Are the calculated bond angles in agreement (see Table 1)? _____

11. Draw the Lewis structure for H_2O.

     a. How many electron domains does the oxygen atom have in H_2O? _____

     b. Which electron domain geometry in Figure 1 applies to H_2O? _____

     c. Are the calculated bond angles in agreement (see Table 1)? _____

12. Draw the Lewis structure for NO_3^-.

     a. How many electron domains does the nitrogen atom have in NO_3^-? _____

     b. Which electron domain geometry in Figure 1 applies to NO_3^-? _____

     c. Are the calculated bond angles in agreement (see Table 1)? _____

13. Draw the Lewis structure for CO_2.

     a. How many electron domains does the carbon atom have in CO_2? _____

     b. Which electron domain geometry in Figure 1 applies to CO_2? _____

     c. Are the calculated bond angles in agreement (see Table 1)? _____

Information:

The names for molecular shapes (molecular geometry) are based on the *arrangements of the **atoms*** in the molecule—not on the arrangements of the electron domains.

Figure 2. The electron geometry, molecular geometry, and Lewis structure of H_2O.

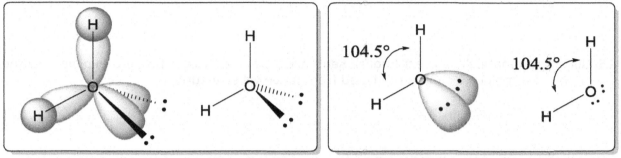

Tetrahedral Electron Geometry   **Bent Molecular Geometry**

The electron geometry of the water molecule is *tetrahedral*, because there are four electron domains (two bonding domains and two nonbonding domains). The molecular geometry of the water molecule is said to be *bent* because the three atoms are not in a straight line. The two lone pairs are not included in the molecular geometry, but they push the two bonding pairs a little closer together than the ideal 109.5°. The actual bond angle, determined by experiment, is 104.5°.

Figure 3. Five common molecular shapes.

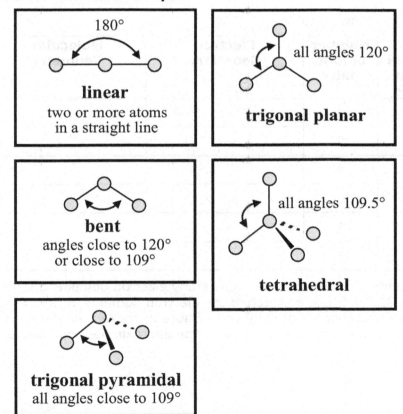

Critical Thinking Questions:

14. Considering the geometries defined in Figure 1, discuss with your team and explain why there are two choices for the bond angle in bent molecules—close to 109° or close to 120°.

15. Using grammatically correct English sentences, work with your team to explain how the shape of a molecule can be predicted from its Lewis structure.

16. List any remaining questions about predicting shapes and angles for molecules.

17. List one strength of your team work today and why it helped your understanding.

Exercises:

1. Complete the following table.

Total number of electron domains	Number of lone pairs	Electron Geometry	Molecular Geometry	Ideal bond angles
2	0			
3	0			
3	1			
4	0			
4	1			
4	2			

2. Draw the Lewis structure for NH_3. How many electron domains does the nitrogen atom have in NH_3? _____ Make a sketch of the electron domains in NH_3. Examine the drawing for the molecular shape of H_2O given in Figure 2; make a similar drawing for NH_3. Name the shape of the NH_3 molecule, and give the approximate bond angles.

3. Draw the Lewis structure for CH_4. How many electron domains does the carbon atom have in CH_4? _____ Make a sketch of the electron domains in CH_4. Examine the drawing for the molecular shape of H_2O given in Figure 2; make a similar drawing for CH_4. Name the shape of the CH_4 molecule, and give the approximate bond angles.

4. Draw the Lewis structure for SO_2. How many electron domains does the sulfur atom have in SO_2? _____ Make a sketch of the electron domains in SO_2. Examine the drawing for the molecular shape of H_2O given in Figure 2; note that the electron domain geometry is different for SO_2, and make a similar drawing for SO_2. Name the shape of the SO_2 molecule, and give the approximate bond angles.

5. On your own paper, draw the Lewis structure, sketch the molecules, predict the molecular shape, and give the bond angles for: PH_3, CO_2, SO_3^{2-}, H_3O^+, NH_2F, H_2CO. The central atom is the first atom listed (other than hydrogen).

6. Predict the bond angles around each atom designated with an arrow in the amino acid glycine, shown below.

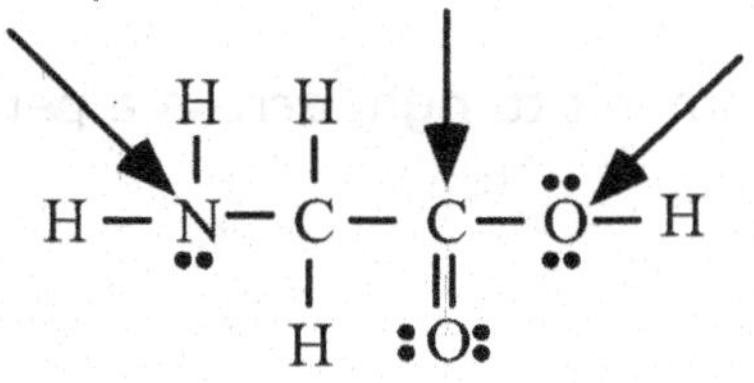

7. Read the assigned pages in your textbook and work the assigned problems.

Polar and Nonpolar Covalent Bonds
(Are atoms in molecules completely neutral?)

Model 1: Relative electronegativity values for selected elements

Linus Pauling noticed that bonds between *different* elements ("heteronuclear") appeared to be stronger than bonds between two atoms of the *same* element ("homonuclear"). He proposed that the bonding electrons in heteronuclear molecules were not shared equally, reasoning that one atom attracted the electrons in the bond more strongly than the other atom.

The **electronegativity** of an atom is its ability to *attract electrons in a covalent bond* closer to itself. Fluorine is the most electronegative element.

Table 1. Pauling's electronegativities for selected elements.

H 2.20							
Li 0.98	Be 1.57		B 2.04	C 2.55	N 3.04	O 3.44	F 3.98
Na 0.93	Mg 1.31		Al 1.61	Si 1.90	P 2.19	S 2.58	Cl 3.16
K 0.82	Ca 1.00	Sc 1.36	Ga 1.81	Ge 2.01	As 2.18	Se 2.55	Br 2.96
Rb 0.82	Sr 0.95	Y 1.22	In 1.78	Sn 1.96	Sb 2.05	Te 2.1	I 2.66

Critical Thinking Questions:

1. Describe the trend in electronegativity of elements from left to right across a period of the Periodic Table.

2. Describe the trend in electronegativity from top to bottom of a group of the Periodic Table.

3. When carbon and oxygen are covalently bonded, are the electrons in the bond attracted closer to **carbon** or **oxygen** (circle one)?

4. Which atom (C or O) is at the *negatively* polarized end of a bond between carbon and oxygen?_____ Which atom is at the positively polarized end? _____ Write a team explanation of your answer in a sentence.

5. Using the data in Model 1, determine which atom in a bond between *carbon* and *nitrogen* would be negatively polarized (C or N). As a team, explain your answer.

Model 2: Types of bonds based on difference in electronegativity

Bond between	Chemical Formula	Electronegativity difference	Type of bond
H and H	H_2	0	nonpolar, covalent
F and F	F_2	0	nonpolar, covalent
C and H	CH_4	0.25	*slightly* polar, covalent
C and O	CO_2	0.89	polar, covalent
H and F	HF	1.78	polar, covalent
Li and F	LiF	3.00	ionic

Critical Thinking Questions:

6. Model 2 states that a H–H bond is **nonpolar**. Devise a team definition for the term *nonpolar*.

7. Considering Model 2, which would you expect to be *more* polar—a C–O bond, or an H–F bond? Discuss with your team and write a consensus explanation.

   For the bond you chose, write the word "*very*" in front of "polar, covalent" in the "type of bond" column in Model 2.

8. Which would be **more** polar: a C-O bond, or a C-Cl bond? Why?

9. The Greek letter delta, δ, is often used to mean "slightly" or "partially." So, $\delta-$ would mean "partially negative," and $\delta+$ would mean "partially positive." For each atom in the molecules below that would have a partial charge, place the symbols $\delta-$ or $\delta+$ near the atom to indicate its charge.

CA17

10. Based on the difference in their electronegativities, a bond between lithium and fluorine would be extremely polarized (see Model 2). Which end of the bond would be *negatively* polarized, **Li** or **F** (circle one)? As a team, discuss how this is consistent with the guideline learned earlier in the course that the bond between a metal and a nonmetal is ionic. Write a sentence to explain.

11. We normally write the formula for lithium fluoride as a lithium ion (Li^+) and a fluoride ion (F^-). Why is this more correct that writing them as "$Li^{\delta+}$" and "$F^{\delta-}$"? Does your team agree?

12. Explain the following statement: The ionic character of a bond increases as the electronegativity difference between the two bonded atoms increases.

13. A rule of thumb says that the closer an element is to fluorine on the periodic table, the more electronegative it is. Based on this rule, place the symbols δ- and δ+ near each atom below that would have a partial positive or negative charge.

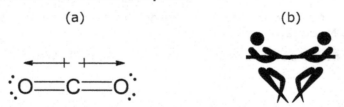

Model 3: Carbon dioxide is overall nonpolar

(a)   (b)

Consider the model of carbon dioxide in (a) above. Both C–O bonds are polar, with the negative ends on each oxygen. The arrows, or vectors, indicate that each electronegative oxygen is pulling electrons equally toward itself. Since both bonds are equally polarized in opposite directions, no particular side or end of the CO_2 molecule is negatively or positively polarized. In other words, the individual bond polarities "cancel out."

This type of canceling can occur whenever **all** the electrons around a central atom are in **identical** covalent bonds (to the same elements) and therefore are equally polarized.

As an analogy, consider image (b) in Model 3. Picture the C–O bonds as ropes, with the two oxygens pulling on the ropes in a "tug-of-war" game. Since they are "pulling" the electrons equally in opposite directions, there is no net movement of the electrons. So even though each C–O bond is polar, the molecule as a whole is **nonpolar**.

Critical Thinking Questions:

14. Place the symbols δ- or δ+ near each atom in Model 3(a) that would have a partial positive or negative charge.

15. Three of the four molecules in CTQ 9 have polar *bonds*, but only one *molecule* is polar. Write a team consensus explanation.

16. One molecule in CTQ 13 is *nonpolar*. Circle it. Explain why it is overall nonpolar.

17. Discuss with your team to arrive at a rule for determining whether a particular <u>bond</u> in a covalent molecule is polar. Write your rule in a complete sentence or two.

18. Discuss with your team to arrive at a rule for determining whether a particular molecule <u>as a whole</u> is polar or nonpolar. Write your rule in a complete sentence or two.

19. What did you encounter today that is confusing to one or more of your team members?

Exercises:

1. Complete the table below. The first one has been done as an example.

Formula	Total valence electrons	Lewis structure	Shape around most central atom	Approximate bond angles around most central atom	Polar molecule or nonpolar molecule
CF_4	32	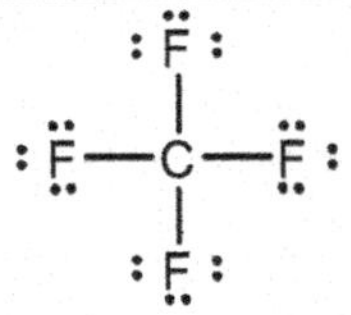 :F: :F—C—F: :F:	tetrahedral	109.5°	nonpolar molecule
PCl_3					
CH_2O (C is central)					
$CHCl_3$ (C is central)					
CH_3OH (C is central, bond the last H to the O)					
HCN					
OF_2					

2. Go back to your Lewis structures in Exercise 1 and indicate the bond polarities by placing δ− or δ+ near all the atoms. Exception: The polarity of a C-H bond is so slight that it is normally ignored.

3. Read the assigned pages in your textbook and work the assigned problems.

The Mole Concept[*]
(What is a mole?)

Model 1: Some selected conversion factors and information

1 dozen items = 12 items

1 score of items = 20 items

1 myriad items = 10,000 items

1 mole of items= 6.022×10^{23} items (Avogadro's Number)

One elephant has one trunk and four legs.

One methane molecule, CH_4, contains one carbon atom and four hydrogen atoms.

Critical Thinking Questions:

Manager: Rotate team members to offer the first explanation for CTQs 1 – 4.

1. How many trunks are found in one dozen elephants? Write your answer as a <u>number</u> and a <u>unit</u>. (The unit should be "trunks.")

2. How many legs are found in one dozen elephants? Write a number and a unit.

3. How many trunks are found in one score of elephants? Write a number and a unit.

4. How many legs are found in one myriad elephants? Write a number and a unit.

Continue writing units with all numerical answers.

5. How many trunks are found in one mole of elephants?

6. How many carbon atoms are found in one dozen methane molecules?

7. How many hydrogen atoms are found in one myriad methane molecules?

8. a. How many elephants are there in one mole of elephants?

   b. Write the answer from part (a) as a conversion factor, with units. Then write a second conversion factor that is the reciprocal of the first conversion factor.

---

[*] Adapted from ChemActivity 28, Moog, R.S. and Farrell, J.J. *Chemistry: A Guided Inquiry*, 5th ed., Wiley, 2011, pp. 164-166.

CA18

Manager: *Ask if any teammates need further clarification from the instructor about the previous questions, and if so, communicate this to the instructor. Ensure that all team members proceed with CTQ 9 (and following) together.*

9. a. How many legs are found in one mole of elephants?

    b. Write a sentence to describe how you calculated the answer in part (a).

10. a. How many trunks are found in one-half mole of elephants?

    b. Write a sentence to describe how you calculated the answer in part (a).

11. How many carbon atoms are found in one mole of methane molecules? Does your team agree?

12. How many hydrogen atoms are found in one-half mole of methane molecules?

13. How many methane molecules are there in one mole of methane?

Model 2: The relationship between average atomic mass and moles

The mass of <u>1 mole</u> of any pure substance is equal to the <u>average atomic mass</u> of that substance <u>expressed in grams</u>.

For example: The average atomic mass of a helium atom is 4.003 amu (atomic mass units).

Therefore: 4.003 g He = 1 mol He

Critical Thinking Questions:

Manager: *Rotate team members to offer the first explanation for CTQs 14 – 17.*

14. Look at a periodic table. What is the average mass (in amu) of one carbon atom?

15. What is the mass (in grams) of one mole of carbon atoms?

16. a. What is the average mass (in amu) of one methane molecule?

    b. Write a sentence to describe how you calculated the answer in part (a).

17. What is the mass (in grams) of one mole of methane molecules?

18. Work with your team to write a grammatically correct English sentence to describe how the mass in amu of one molecule of a compound is related to the mass in grams of one mole of that compound.

19. As a team, identify two main concepts from this ChemActivity.

Exercises:

Unless otherwise stated, calculate all mass values in grams.

1. Consider CTQ 1. A unit plan for this problem would be dozens → elephants → trunks. Since you know there are <u>12 elephants in 1 dozen elephants</u> and <u>1 trunk per elephant</u>, you could set up the problem using conversion factors as follows:

$$1 \text{ dozen elephants} \times \frac{12 \text{ elephants}}{1 \text{ dozen elephants}} \times \frac{1 \text{ trunk}}{1 \text{ elephant}} = 12 \text{ trunks}$$

On your own paper, complete similar unit conversions, showing all work, for CTQs 2-13.

2. If you weigh out 69.236 g of lead, how many atoms of lead do you have? Make a unit plan first. Show work.

3. Consider 1.00 mole of dihydrogen gas, H_2. How many dihydrogen molecules are present? How many hydrogen atoms are present? What is the mass of this sample?

4. Ethanol has a molecular formula of CH_3CH_2OH.
   a. What is the average mass of one molecule of ethanol?

   b. What is the mass of 1.000 moles of ethanol?

   c. What is the mass of 0.5623 moles of ethanol, CH_3CH_2OH?

   d. How many moles of ethanol are present in a 100.0 g sample of ethanol?

e. How many moles of each element (C, H, O) are present in a 100.0 g sample of ethanol?

f. How many grams of each element (C, H, O) are present in a 100.0 g sample of ethanol?

5. How many moles of carbon dioxide, CO_2, are present in a sample of carbon dioxide with a mass of 254 grams? How many moles of O atoms are present?

6. Indicate whether each of the following statements is true or false, and explain your reasoning.
    a. One mole of NH_3 weighs more than one mole of H_2O.

    b. There are more carbon atoms in 48 grams of CO_2 than in 12 grams of diamond (a form of pure carbon).

    c. There are equal numbers of nitrogen atoms in one mole of NH_3 and one mole of N_2.

    d. The number of Cu atoms in 100 grams of Cu(s) is the same as the number of Cu atoms in 100 grams of copper(II) oxide, CuO.

    e. The number of Ni atoms in 100 moles of Ni(s) is the same as the number of Ni atoms in 100 moles of nickel(II) chloride, $NiCl_2$.

    f. There are more hydrogen atoms in 2 moles of NH_3 than in 2 moles of CH_4.

7. Read the assigned pages in your text, and work the assigned problems.

Balancing Chemical Equations
(What stays the same and what may change in a chemical reaction?)

Model: Atoms are conserved in chemical reactions

Atoms are neither created nor destroyed when chemical reactions occur. Two balanced chemical reactions (or equations) are given below. In a chemical reaction equation, reactant atoms are shown on the left side of the arrow and product atoms are shown on the right.

Equation (1): $2 H_2 (g) + O_2 (g) \longrightarrow 2 H_2O (g)$

Molecular Picture:

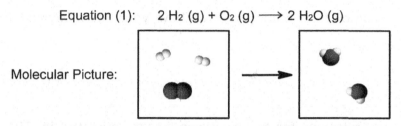

Equation (2): $Fe_2O_3 (s) + 2 Al (s) \longrightarrow 2 Fe (l) + Al_2O_3 (s)$

Critical Thinking Questions:

1. Considering Equation (1) above:
   a. What are the reactants? Write your answer in both words and formulas. Check that your teammates agree.

   b. What is the product? Write your team's answer in both words and formulas.

2. What does the arrow represent in a chemical reaction?

3. How many hydrogen <u>molecules</u> are represented on the reactant side of equation (1)? ___

4. How many water <u>molecules</u> are represented on the product side of equation (1)? ___

5. How many H <u>atoms</u> are represented on the reactant side of equation (1)? _____
   On the product side? _____

6. How many O atoms are represented on the reactant side of equation (1)? _____
   On the product side? _____

7. Compare the number of <u>atoms</u> of each element (H, O) on the reactant and product sides of equation (1). Are these numbers **the same** or **different** (circle one)?

8. Compare the number of <u>atoms</u> of each element (Fe, Al, O) on the reactant and product sides of equation (2). Are these numbers **the same** or **different** (circle one)?

9. In reaction (1):
   a. For each H_2 <u>molecule</u> consumed, how many H_2O <u>molecules</u> are produced? ___
   b. For each O_2 molecule consumed, how many H_2O molecules are produced? ___
   c. Recall that a mole is simply a specific large quantity, like a dozen or a gross. For each <u>mole</u> of oxygen molecules consumed, how many <u>moles</u> of water molecules are produced? ___

Manager: *Ensure that no team members proceed to CTQ 10 until you have checked for agreement on all your answers to CTQs 3-9.*

10. Considering your answers to CTQ 9b and 9c, what two quantities can be indicated by the <u>coefficient</u> (the number) in front of the chemical formulas?

11. For each one mole of H_2O molecules that are produced in reaction (1), how many moles of O_2 molecules are required?

12. Describe in a complete sentence or two how your team arrived at your answer to CTQ 11.

13. One of the interpretations for the coefficients that you listed in CTQ 10 is possible for your answer to CTQ 11, and one is not. Explain.

14. Considering reaction (1), what is the total number of molecules of water produced from two molecules of hydrogen and one molecule of oxygen?_____

15. Is the <u>total number of molecules</u> identical on the reactant and product sides of these balanced equations? _____

16. Explain how your answer to CTQ 15 can be consistent with the idea that atoms are neither created nor destroyed when chemical reactions take place.

17. a. If we know that hydrogen gas burns in oxygen gas to produce water vapor, why can we not simply write the equation as follows? H_2 (g) + O_2 (g) $\longrightarrow$ H_2O (g)

   b. Why can we not "fix" this equation by writing it as follows? H_2 (g) + O (g) $\longrightarrow$ H_2O (g)

   c. Compare the equation in (a) above with equation (1) in the Model. What was done to the equation in the model to "fix" the equation?

   d. Is the molecular picture below a valid depiction of equation (1) in the model? Explain.

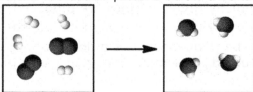

Exercises:

1. When we know what the reactants and products are for a chemical reaction, we can determine their ratios by balancing the numbers of atoms on the reactant and product sides of the equation. We do this by adding coefficients in front of each reactant and product, using a process of trial and error. Add coefficients to balance the following equations.

    a. $CH_4\ (g) +\quad O_2\ (g) \rightarrow\quad CO_2\ (g) +\quad H_2O\ (g)$

    b. $Fe_3O_4\ (l) +\quad CO\ (g) \rightarrow\quad FeO\ (l) +\quad CO_2\ (g)$

    c. $Cr\ (s) +\quad S_8\ (s) \rightarrow\quad Cr_2S_3\ (s)$

2. For each mole of S_8 molecules consumed in the reaction in Exercise 1c, how many moles of Cr_2S_3 molecules are produced?

3. In the balanced reaction: $2\ CO\ (g) + O_2\ (g) \rightarrow 2\ CO_2\ (g)$

    a. How many moles of CO_2 can be produced from 4 moles CO and 2 moles O_2?

    b. How many grams of CO_2 would this be?

4. Add molecular (spacefilling) pictures to the boxes below in the correct ratio so that it represents reaction (1) in the Model. For example, ● could be used to represent one molecule of H_2.

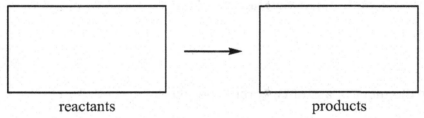

reactants   products

5. Use molecular (spacefilling) pictures in the correct ratio to represent the reaction in Exercise 1a. Be sure that the representations for C atoms, H atoms and O atoms are distinguishable. (You may wish to include a key.)

6. Read the assigned pages in your text, and work the assigned problems.

Stoichiometry
(How are masses of reactants and products related?)

Model 1: The Balanced Chemical Equation

As we saw in ChemActivity 19, a balanced chemical reaction can be interpreted in two ways. First, it can describe how many molecules of reactants are consumed in order to produce a certain number of molecules of products. Second, it can describe how many *moles* of reactants are consumed in order to produce the indicated number of *moles* of products.

$$3 H_2 (g) \ + \ N_2 (g) \ \rightarrow \ 2 NH_3 (g) \qquad\qquad (1)$$

$$2 CO (g) \ + \ O_2 (g) \ \rightarrow \ 2 CO_2 (g) \qquad\qquad (2)$$

Critical Thinking Questions:

1. How many NH_3 molecules are produced for every N_2 molecule that is consumed in reaction (1)?

2. How many moles of N_2 react to produce 8 moles of NH_3 in reaction (1)? _____ mol N_2

3. Discuss with your team, and describe in a sentence how you calculated the answer to CTQ 2.

4. Show your work for CTQ 2 as a unit conversion. Make sure that your conversion factor contains units.

5. How many moles of CO_2 will be formed when 0.25 mol of O_2 react in reaction (2)? Show your work, including a conversion factor.

6. For reaction (2):
   a. What is the molar mass of O_2 (grams per mole)?

   b. How many moles of O_2 will react when starting with 10.0 g of O_2? Show your work, with a conversion factor.

   c. How many moles of CO_2 will be produced when the moles of O_2 calculated in (b) are consumed? Show work.

   d. How many grams of CO_2 will be produced from the moles of CO_2 calculated in (c)?

7. Consult with your team and write a list of steps needed to calculate the number of grams of CO_2 produced in CTQ 6.

8. For the reaction: $2\ C_2H_2 + 5\ O_2 \rightarrow 4\ CO_2 + 2\ H_2O$

   How many grams of carbon dioxide (CO_2) are produced when 22.0 g of C_2H_2 are burned? Work with your team to find a consensus answer. *Manager:* Before continuing, ensure that all team members can follow the steps in CTQ 7 to solve this problem.

Model 2: Percent yield

A researcher performed the reaction in CTQ 6, but only 5.20 g of CO_2 was collected.

The amount you calculated in CTQ 6 is what the reaction can theoretically produce. But the researcher only collected 5.20 g. Perhaps some gas escaped, or perhaps the reaction was incomplete. The **percent yield** is:

$$\frac{\text{amount (g) produced}}{\text{amount (g) theoretical}} \times 100 = \%\ \text{yield}$$

Critical Thinking Questions:

9. Calculate the percent yield for the reaction in CTQ 6 given the amount the researcher collected.

10. Methane gas (CH_4) burns in oxygen (O_2) to produce carbon dioxide and water. Work with your team to answer the following questions.
    a. Write a balanced equation for the reaction. (Phase labels are not necessary.)

    b. How many grams of carbon dioxide could be produced from 2.0 g of O_2?

    c. What is the percent yield if only 1.0 gram of carbon dioxide is collected?

11. What are one or two important concepts in this activity?

CA20A

12. List one or two problem solving strategies that your team utilized today.

Exercises:

1. For the reaction: ___ Al + ___ Br_2 → ___ $AlBr_3$
   a. Balance the reaction.
   b. How many grams of $AlBr_3$ may be produced from 10.0 g of bromine?

   c. What is the percent yield if only 3.2 grams were obtained?

2. For the reaction: ___ CS_2 + ___ NH_3 → ___ H_2S + ___ NH_4SCN
   a. Balance the reaction.
   b. How many grams of hydrogen sulfide (H_2S) are produced when 0.0365 grams of carbon disulfide are consumed?

   c. What is the percent yield if only 0.0061 grams were obtained?

3. For the reaction: ___ Fe_2S_3 + ___ HCl → ___ $FeCl_3$ + ___ H_2S
   a. Balance the above reaction
   b. How many grams of iron(III) chloride are produced when 26.0 grams of hydrogen sulfide gas are produced?

   c. What is the percent yield if only 26.4 grams were obtained?

4. (Answer on a separate piece of paper.) The thermite reaction has been used for welding railroad rails, in incendiary bombs, and to ignite solid-fuel rockets. The reaction is:

   $$Fe_2O_3 \text{ (s) } + 2 \text{ Al (s) } → 2 \text{ Fe (l) } + Al_2O_3 \text{ (s)}$$

   What masses of iron(III) oxide and aluminum must be used to produce 25.0 g of iron? What is the mass of aluminum oxide that would be produced?

5. Read the assigned pages in your text, and work the assigned problems.

Limiting Reagent[*]
(How much can you make?)

Model 1: The S'more:

A delicious treat known as a S'more is constructed with the following ingredients and amounts:

> 1 graham cracker
> 1 chocolate bar
> 2 marshmallows

At a particular store, these items can be obtained only in full boxes, each of which contains one gross of items. A gross is a specific number of items, analogous (but not equal) to one dozen. The boxes of items have the following net weights (the weight of the material inside the box):

> box of graham crackers   9.0 pounds
> box of chocolate bars   36.0 pounds
> box of marshmallows   3.0 pounds

Critical Thinking Questions:

1. If you have a collection of 100 graham crackers, how many chocolate bars and how many marshmallows do you need to make S'mores using all the graham crackers?

2. If you have a collection of 1000 graham crackers, 800 chocolate bars, and 1000 marshmallows:

   a. How many S'mores can you make?

   b. What (if anything) will be left over, and how many of that item will there be?

Information:

Chemists refer to the reactant which limits the amount of product which can be made from a given collection of original reactants as the **limiting reagent** or **limiting reactant.**

---

[*] Adapted from ChemActivity 30, Moog, R.S. and Farrell, J.J. *Chemistry: A Guided Inquiry.* Wiley, 5th ed., 2011, pp. 174-177.

Critical Thinking Questions:

3. Identify the limiting reagent in CTQ 2. Does your team agree?

4. Based on the information given:

   a. Which of the three ingredients (a graham cracker, a chocolate bar, or a marshmallow) weighs the most?

   b. Which weighs the least?

   Discuss as a team, and explain your reasoning.

5. If you have 36.0 pounds of graham crackers, 36.0 pounds of chocolate bars, and 36.0 pounds of marshmallows:

   a. Which item do you have the most of?

   b. Which item do you have the least of?

   Explain your reasoning.

6. If you attempt to make S'mores from the material described in CTQ 5:

   a. What will the limiting reagent be?

   b. How many gross of S'mores will you have made?

   c. How many gross of each of the two leftover items will you have?

   d. How many pounds of each of the leftover items will you have?

   e. How many pounds of S'mores will you have?

7. Using G as the symbol for graham cracker, Ch for chocolate bar, and M for marshmallow, write a "balanced chemical equation" for the production of S'mores. Compare your answer with your team.

8. Work with your team and explain in complete sentences why it is not correct to state that if you start with 36 pounds each of G, Ch, and M then you should end up with 108 (3 x 36 = 108) pounds of S'mores.

Model 2: Water

Water can be made by burning hydrogen in oxygen, as shown below:

$$2 H_2 + O_2 \rightarrow 2 H_2O$$

A mole is a specific number of items, analogous (but not equal) to one dozen. The reactants in the equation have the following molar masses (the weight of the material in a mole of material):

mole of H_2 molecules   2.0 grams
mole of O_2 molecules   32.0 grams

Critical Thinking Questions:

9. Based on the information given:

   a. Which of the two ingredients (hydrogen molecules or oxygen molecules) weighs the most?

   b. Which weighs the least?

   As a team, explain your reasoning.

10. If you have a collection of 100 hydrogen molecules, how many oxygen molecules do you need to make water with all of the hydrogen molecules?

11. If you have 32.0 grams of hydrogen molecules and 32.0 grams of oxygen molecules,

   a. Which item do you have the most of?

   b. Which item do you have the least of?

   Explain your reasoning.

Manager: *Ensure that your team works together to answer the following questions.*

12. If you attempt to make water from the material described in CTQ 11:

   a. What is the limiting reagent?

   b. How many moles of water molecules will you have made?

   c. How much does one mole of water molecules weigh? Explain how you can determine this from Model 2.

   d. Based on your answer to (b), how many grams of water will you have?

   e. A student performed the reaction in CTQ 11 and obtained 34.0 grams of water. What percentage of the total possible amount of water did he obtain?

13. List one or two problem solving strategies that your team developed today.

14. List one improvement your team could make for next time and how you intend to implement it.

Exercises:

1. The thermite reaction, once used for welding railroad rails, is often used for an exciting chemistry demonstration because it produces red-hot molten iron. The reaction is:

$$Fe_2O_3 (s) + 2 Al (s) \rightarrow 2 Fe (l) + Al_2O_3 (s)$$

If you start with 50.0 g of iron(III) oxide and 25.0 g of aluminum, what is the limiting reagent? What is the maximum mass of aluminum oxide that could be produced? How much aluminum oxide would be produced if the yield is 93%?

2. How much aspirin can be made from 100.0 g of salicylic acid and 100.0 g of acetic anhydride? If 122 g of aspirin are obtained by the reaction, what is the percent yield?

$C_7H_6O_3$	+	$C_4H_6O_3$	$\rightarrow$	$C_9H_8O_4$	+	$HC_2H_3O_2$
salicylic acid		acetic anhydride		aspirin		acetic acid

3. Read the assigned pages in your text, and work the assigned problems.

Predicting Binary Reactions
(When will precipitation or neutralization reactions occur?)

Critical Thinking Question:

1. Write the chemical formula for each compound.

    a. barium nitrate   _____

    b. sodium acetate   _____

    c. magnesium chloride   _____

    d. sodium hydroxide   _____

    e. calcium phosphate   _____

    f. lead(II) sulfate   _____

    g. sulfuric acid   _____

    h. ammonium carbonate   _____

Model 1: Solubility rules for ionic compounds

1. A compound containing a cation from Group 1 (alkali metal) or the ammonium ion (NH_4^+) is likely soluble.
2. A compound containing an anion from group 17 (halides) is likely soluble.
    *   **Exceptions:** Ag^+, Hg_2^{2+}, and Pb^{2+} halides are insoluble.
3. A compound containing nitrate (NO_3^-) or acetate ($CH_3CO_2^-$) is likely soluble.
4. A compound containing sulfate (SO_4^{2-}) is likely soluble.
    *   **Exceptions:** The sulfates of Ba^{2+}, Hg_2^{2+}, and Pb^{2+} are insoluble.
5. **Any other** ionic compounds are likely **insoluble**.

Note: You need not memorize the information in Table 1 unless your instructor requires it.

Critical Thinking Questions:

2. Discuss with your team to recall and record the meanings of the phase labels (aq), (s) and (l).

3. Suppose that each compound in CTQ 1 is placed in water. Examine each rule in Model 1 **in order**. When you reach a rule that applies to the compound, place the symbol (aq) or (s) after each formula to indicate its solubility in water.

Information: Precipitation and neutralization reactions

Recall that binary compounds that dissociate into ions when dissolved in water are referred to as electrolytes. Chemical reactions between binary compounds occur whenever two electrolytes can combine to form an insoluble solid or a covalent molecule such as water.

⇒ Neutralization: acid + base → salt + water
⇒ Precipitation: formation of an insoluble solid (precipitate)

Model 2: Two binary reactions

$$AgNO_3 \text{ (aq)} + NaCl \text{ (aq)} \rightarrow AgCl \text{ (s)} + NaNO_3 \text{ (aq)} \qquad (1)$$
$$Ca(OH)_2 \text{ (aq)} + 2 HCl \text{ (aq)} \rightarrow CaCl_2 \text{ (aq)} + 2 H_2O \text{ (l)} \qquad (2)$$

Critical Thinking Questions:

4. Consider Model 2. What insoluble solid is formed in reaction (1)? _____

5. Consider the information on the previous page. What type of reaction is reaction (1)?

6. What covalent molecule is formed in reaction (2)? _____ What type of reaction is reaction (2)?

7. What two **cations** are present in the reactants in reaction (1)?

8. What two **anions** are present in the reactants in reaction (1)?

9. What two cations are products in reaction (1)?

10. What two anions are products in reaction (1)?

11. With your team, discuss the changes that the silver ion undergoes as reaction (1) occurs. Write your consensus explanation of what happens to the silver ion in a complete sentence.

12. Discuss with your team and write a sentence to explain what changes the nitrate ion undergoes as reaction (1) occurs.

13. Work as a team to compose a list of steps needed to predict the products of a precipitation or neutralization reaction. **Manager:** *Give each member an opportunity to contribute to the list.*

14. a. Write formulas for the two products that are possible when silver nitrate, AgNO₃, reacts with calcium bromide, CaBr₂.

    b. Write and balance a chemical reaction for the solution in (a).

15. For the reaction:

$$2\ KCl\ +\ Pb(NO_3)_2\ \rightarrow\ PbCl_2\ +\ 2\ KNO_3$$

    Write in (aq), (s), or (l) for each compound.

16. Discuss with your team whether any additions or corrections are necessary for your list of steps outlined in CTQ 13, including how to add the phase labels (aq), (s), or (l).

Using your list of steps for writing binary reactions, write a balanced molecular equation starting with each set of reactants below. Write NR if no reaction will occur.

17. Aqueous lead(II) nitrate and aqueous sodium iodide.

18. Aqueous ammonium chloride and aqueous potassium nitrate.

19. Aqueous sodium hydroxide and aqueous sulfuric acid.

Model 3: Writing Net Ionic Equations (optional)

Net ionic equations are used to more accurately depict the reaction of ions to form a product. Ions that do not change and are not involved in the formation of an insoluble solid or covalent molecule are called **spectator ions** and are not included in the net ionic reaction.

Equation: $2\ AgNO_3(aq)\ +\ CaBr_2(aq)\ \rightarrow\ Ca(NO_3)_2(aq)\ +\ 2\ AgBr(s)$

Ionic Equation:
$2\ Ag^+(aq)\ +\ 2\ NO_3^-(aq)\ +\ Ca^{2+}(aq)\ +\ 2\ Br^-(aq)\ \rightarrow\ Ca^{2+}(aq)\ +\ 2\ NO_3^-(aq)\ +\ 2\ AgBr(s)$

Net Ionic Equation: $2\ Ag^+(aq)\ +\ 2\ Br^-(aq)\ \rightarrow\ 2\ AgBr(s)$
Or: $Ag^+(aq)\ +\ Br^-(aq)\ \rightarrow\ AgBr(s)$

Critical Thinking Questions:

20. What is the product of the net ionic equation in Model 3? _____

21. What are the spectator ions of the reaction in Model 3?

22. For the reaction: $BaCl_2(aq)\ +\ K_2SO_4(aq)\ \rightarrow\ BaSO_4(s)\ +\ 2\ KCl(aq)$
    a. Write the ionic equation.

    b. What are the spectator ions?

    c. Write the net ionic equation.

23. For the reaction: $Pb(NO_3)_2(aq) + 2 KI(aq) \rightarrow 2 KNO_3(aq) + PbI_2(s)$
    a. Write the ionic equation.

    b. What are the spectator ions?

    c. Write the net ionic equation.

24. Indicate the major concepts of this activity.

25. What is one area in which your team could use improvement? How can it be achieved next time?

Exercises:

1. Using the solubility rules, write a balanced molecular equation starting with each set of reactants below. Include the phase labels.
    a. Aqueous silver nitrate and aqueous sodium phosphate (Note: silver always has a +1 charge in ionic compounds).

    b. Aqueous calcium nitrate and aqueous potassium carbonate.

    c. Aqueous copper(II) nitrate and aqueous potassium hydroxide.

    d. Aqueous zinc chloride and aqueous sodium sulfide (Note: zinc always has a +2 charge in ionic compounds).

    e. Aqueous lead(II) nitrate and aqueous sodium phosphate.

    f. Aqueous magnesium sulfate and aqueous sodium hydroxide.

2. (optional) Write the net ionic equation for each reaction in Exercise 1.
3. Read the assigned pages in your text, and work the assigned problems.

Oxidation-Reduction Reactions
(What happens when electrons are transferred between atoms?)

Model 1: Oxidation Numbers

Oxidation-reduction reactions (often called "redox" reactions) are those in which *electrons are transferred between atoms*. Oxidation and reduction always occur together, since an atom that loses electrons must give them to an atom that gains the electrons.

A way to keep track of electrons in chemical reactions is by assigning **oxidation numbers** to each element in a reaction. The oxidation number is the charge an atom in a substance would have if the electron pairs in each covalent bond belonged to the more electronegative atom.

An atom that...	is said to be...	and its oxidation number...
gains electrons	reduced	decreases
loses electrons	oxidized	increases

You are responsible for <u>recognizing</u> (but not <u>predicting</u>) redox reactions.

Table 1. General rules for assigning oxidation numbers

1. Pure **elements** (not bonded to other elements) have an oxidation number of 0. This includes diatomic elements like H_2 or O_2.

2. The oxidation number of an **ion** equals its charge. For example:
   - group 1 (alkali metals) $\rightarrow$ +1
   - group 2 (alkaline earth) $\rightarrow$ +2

3. In **molecules, F** is –1; **Cl, Br, I** are –1 if bonded to less electronegative atom

4. **Oxygen** is nearly always –2

5. **Hydrogen** nearly always +1

6. Figure any others by calculation. <u>The sum of oxidation numbers of the atoms of a molecule equals zero; the sum for an ion (including polyatomics) should equal its overall charge</u>.
   - For example, in CO_2, the oxidation number of each O is –2, so the oxidation number of C must be +4.
   - For PO_4^{3-}, the oxidation number of each O is -2 and the total must be -3, so the oxidation number for P is +5.

Critical Thinking Questions:

1. Work as a team to assign an oxidation number to <u>each atom individually</u> in the following substances. The first one has been done for you.

   a. CCl_4 C _+4_   Cl _–1_

   b. CH_4 C ___   H ___

   c. HNO_3 H ___   N ___   O ___

   d. $Ca(NO_3)_2$ Ca ___   N ___   O ___

   e. $K_2Cr_2O_7$ K ___   Cr ___   O ___

   f. SO_4^{2-} S ___   O ___

2. Identify any aspects of assigning oxidation numbers that remain unclear.

Model 2: Recognizing redox reactions.

$$2\ Fe(s)\ +\ 3\ O_2(g)\ \rightarrow\ Fe_2O_3(s) \hspace{3cm} (1)$$

Reaction (1) above is an example of a redox reaction. Both reactants have an oxidation number of zero. In Fe_2O_3, iron has an oxidation number of +3, and oxygen has an oxidation number of -2. So iron changes from 0 to +3, and oxygen goes from 0 to -2. In this example, iron is oxidized, and it acts as the **reducing agent** (causing the reduction of another species). Oxygen is reduced and acts as the **oxidizing agent** (causing the oxidation of another species). Oxidation and reduction always occur in the same reaction.

$$Cu(s)\ +\ 2\ AgNO_3\ (aq)\ \rightarrow\ 2\ Ag(s)\ +\ Cu(NO_3)_2(aq) \hspace{1cm} (2)$$

$$Pb(NO_3)_2(aq)\ +\ 2\ KI(aq)\ \rightarrow\ 2\ KNO_3(aq)\ +\ PbI_2(s) \hspace{1cm} (3)$$

Critical Thinking Questions:

3. For reaction (2) in Model 2, indicate the oxidation numbers for:

	Reactant side	Product side
Cu		
Ag		
N		
O		

4. a. For reaction (2), which atom is oxidized? _____

    b. Which atom is reduced? _____

    c. Which atom or compound acts as the oxidizing agent? _____

    d. Which atom or compound acts as the reducing agent? _____

5. As a team, review reaction (3). Is this a redox reaction? Why or why not?

6. For the reaction: $CH_4(g)\ +\ 2\ O_2(g)\ \rightarrow\ CO_2(g)\ +\ 2\ H_2O(l)$

    a. Is this a redox reaction? If so, identify which atom is oxidized and which atom is reduced.

    b. Which element or compound acts as the oxidizing agent? _____

    c. Which element or compound acts as the reducing agent? _____

7. Explain how one can identify a redox reaction.

Model 3: Redox reactions in biological systems (organic molecules)

Assigning oxidation numbers in complex biological molecules is often difficult, but there is a shortcut. In these cases, a molecule that undergoes **addition of oxygen atoms** or **loss of hydrogen atoms** is **oxidized**. If oxygen atoms are **removed** (*i. e.,* a product) or hydrogen atoms are **added** (*i. e.,* a reactant), the molecule is **reduced**.

Often, we do not balance these reactions, so *the hydrogen or oxygen being added or lost may not be shown*.

The following reactions focus on oxidation or reduction of carbon. Even though the oxidizing or reducing agents are not shown, reduction and oxidation still occur together in these reactions.

Critical Thinking Questions:

8. Working with your team, review the reactions in Model 3 to answer parts (a) – (c).

   a. In reaction (4), is C oxidized or reduced? How can you tell?

   b. In reaction (5), is C oxidized or reduced? How can you tell?

   c. In reaction (6), is C oxidized or reduced? How can you tell?

9. When linoleic acid, an unsaturated fatty acid, reacts with hydrogen, it becomes a saturated fatty acid.

   $$C_{18}H_{32}O_2 \rightarrow C_{18}H_{36}O_2 \text{ (not balanced)}$$

   This process, called hydrogenation, may be used to make shortening ("Crisco") out of vegetable oil. In this hydrogenation, is linoleic acid oxidized or reduced? Consult with your team, and write a consensus explanation.

10. List one or two tips or strategies that your team discovered to help in identifying redox reactions.

11. Identify one strength of your teamwork today, and one area that your team could improve upon.

Exercises:

1. Review and be able to apply the oxidation number rules in Model 1.
2. Review the sections on combustion and respiration reactions in your textbook. For our purposes, the definition of a ***combustion reaction*** is when a molecule (methane, glucose, etc.) reacts with oxygen (O_2) to produce CO_2 and H_2O.
3. Assign oxidation numbers to each atom in the reactions below. Then tell which element is oxidized and which is reduced in each reaction.

   a. $CuO(s) + H_2(g) \rightarrow Cu(s) + H_2O(g)$

   b. $2\ CO(g) + O_2(g) \rightarrow 2\ CO_2(g)$

4. Is the reaction in Exercise 3a an **oxidation** or a **reduction** **(circle one)** of copper(I) oxide? Explain how you can tell **without** assigning oxidation numbers.

5. Is the reaction in Exercise 3b an **oxidation** or **reduction** **(circle one)** of carbon monoxide? Explain how you can tell **without** assigning oxidation numbers.

6. Are the reactions below redox reactions? Explain your answer in terms of oxidation numbers.

   a. $Cu(s) + 2\ AgNO_3(aq) \rightarrow 2\ Ag(s) + Cu(NO_3)_2(aq)$

   b. $Pb(NO_3)_2(aq) + 2\ KI(aq) \rightarrow 2\ KNO_3(aq) + PbI_2(s)$

7. Consider the balanced net ionic equation:

   $$5\ H_2C_2O_4(aq) + 2\ MnO_4^-(aq) + 6\ H^+(aq) \rightarrow 10\ CO_2(g) + 2\ Mn^{2+}(aq) + 8\ H_2O(l)$$

   Is the reaction a redox reaction? Explain your answer in terms of oxidation numbers.

8. When ethanol is ingested and metabolized in the liver, it is first converted to acetaldehyde, and then to acetic acid, as shown in the Lewis structures below.

ethanol — acetaldehyde (ethanal) — acetic acid (ethanoic acid)

   a. Is the conversion of ethanol to acetaldehyde an **oxidation** or **reduction** **(circle one)**? Explain how you can tell.

   b. Is the conversion of acetaldehyde to acetic acid an **oxidation** or **reduction** **(circle one)**? Explain how you can tell.

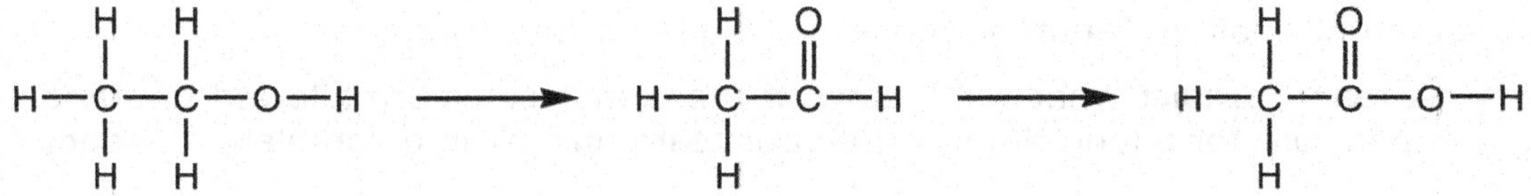

9. Read the assigned pages in your text, and work the assigned problems.

Equilibrium*
(Do reactions really ever stop?)

Model 1: The conversion of *cis*-2-butene to *trans*-2-butene.

Consider a simple chemical reaction where the forward reaction occurs in a single step and the reverse reaction occurs in a single step:

$$A \rightleftharpoons B$$

The following chemical reaction, where *cis*-2-butene is converted into *trans*-2-butene, is an example.

| cis-2-butene | trans-2-butene |

In this example, one end of a *cis*-2-butene molecule rotates 180° to form a *trans*-2-butene molecule. Rotation around a double bond rarely happens at room temperature because the collisions are not sufficiently energetic to weaken the double bond. At higher temperatures, around 400°C for *cis*-2-butene, collisions are sufficiently energetic and an appreciable reaction rate is detected.

Critical Thinking Questions:

1. In most molecular model kits, black = C and white = H. Short bonds are used for single bonds and <u>two</u> of the longer, flexible bonds are used for double bonds.
   a. Have one team member make a model of *cis*-2-butene with a modeling kit.
   b. Have a second team member make a model of *trans*-2-butene.

2. Examine the models as a team. What must be done to convert *cis*-2-butene to *trans*-2-butene?

3. What must be done to convert *trans*-2-butene to *cis*-2-butene?

4. Can the molecules be interconverted without disconnecting ("breaking") the bonds? ____

5. A large number of *cis*-2-butene molecules is placed in a container.

   a. As a team, predict what will happen if these molecules are allowed to stand at room temperature for a long time. Write your team answer in a complete sentence.

   b. As a team, predict what will happen if these molecules are allowed to stand at 400°C for a long time. Write your team answer in a complete sentence.

---

* Adapted from ChemActivity 37, Moog, R.S. and Farrell, J.J. *Chemistry: A Guided Inquiry*, 5th ed., Wiley, 2011, pp. 211-216.

Model 2. The number of molecules as a function of time.

Consider the simple reaction:

$$A \rightleftharpoons B$$

The system is said to be at equilibrium when the concentrations of reactants and products stop changing.

Imagine the following hypothetical system. Exactly 10,000 A molecules are placed in a container which is maintained at 800°C. We have the ability to monitor the number of A molecules and the number of B molecules in the container at all times. We collect the data at various times and compile Table 1.

Table 1. Number of A and B molecules as a function of time

Time (seconds)	Number of A Molecules	Number of B Molecules	Number of A Molecules that React in Next Second	Number of B Molecules that React in Next Second	Number of A Molecules Formed in Next Second	Number of B Molecules Formed in Next Second
0	10000	0	2500	0	0	2500
1	7500	2500	1875	250	250	1875
2	5875	4125	1469	413	413	1469
3	4819	5181	1205	518	518	1205
4	4132	5868	1033	587	587	1033
5	3686	6314	921	631	631	921
6	3396	6604	849	660	660	849
7	3207	6793	802	679	679	802
8	3085	6915	771	692	692	771
9	3005	6995	751	699	699	751
10	2953	7047	738	705	705	738
11	2920	7080	730	708	708	730
12	2898	7102	724	710	710	724
13	2884	7116	721	712	712	721
14	2874	7126	719	713	713	719
15	2868	7132	717	713	713	717
16	2864	7136	716	714	714	716
17	2862	7138	715	714	714	715
18	2860	7140	715	714	714	715
19	2859	7141	715	714	714	715
20	2858	7142	715	714	714	715
21	2858	7142	714	714	714	714
22	2858	7142	714	714	714	714
23	2857	7143	714	714	714	714
24	2857	7143	714	714	714	714
25	2857	7143	714	714	714	714
30	2857	7143	714	714	714	714
40	2857	7143	714	714	714	714
50	2857	7143	714	714	714	714

Critical Thinking Questions:

Refer to Table 1 to answer CTQs 6-12. Note that <u>no calculations are required</u> for CTQs 6-7.

6. During the time interval 0–1 s:
    a. How many A molecules react? _____
    b. How many B molecules are formed? _____
    c. Why are these two numbers equal?

7. During the time interval 10–11 s:
    a. How many B molecules react? _____
    b. How many A molecules are formed? _____
    c. Why are these two numbers equal?

8. a. During the time interval 0–1 s, what fraction of the A molecules react?

    b. During the time interval 10–11 s, what fraction of the A molecules react?

    c. During the time interval 24–25 s, what fraction of the A molecules react?

9. During the time interval 100–101 s, how many molecules of A react? Work with your team to explain your reasoning.

10. a. During the time interval 1–2 s, what fraction of the B molecules react?

    b. During the time interval 10–11 s, what fraction of the B molecules react?

    c. During the time interval 24–25 s, what fraction of the B molecules react?

11. During the time interval 100–101 s, how many molecules of B react? Work with your team to explain your reasoning.

12. For the reaction described in Table 1:

    a. How long did it take for the reaction to come to equilibrium?

    b. Are A molecules still reacting to form B molecules at t = 500 seconds?

    c. Are B molecules still reacting to form A molecules at t = 500 seconds?

Information:

For the process in Model 2, the rate of conversion of B to A equals the number of B molecules that react per second, or:

$$\text{rate of conversion of B to A} = \frac{\Delta \text{ number of B molecules}}{\Delta t}$$

where "Δ" means "change in."

The relationship between the rate of conversion of B to A and the number of B molecules is given by equation (1):

$$\text{rate of conversion of B to A} = (\text{fraction}) \times (\text{number of B molecules}) \qquad (1)$$

where the fraction is a specific value called a *rate constant*, k_B.

Critical Thinking Questions:

13. Rewrite equation (1), replacing "(fraction)" with "k_B."

14. Considering your answers to CTQ 10, what is the value of k_B in equation (1) in molecules per second? Be sure to include the units in your answer.

15. Rewrite equation (1) again, replacing "k_B" with the value of k_B that you determined in CTQ 14.

16. Write a mathematical equation analogous to equation (1) for the rate of conversion of A molecules into B molecules. This equation should include a constant k_A.

17. What is the value of k_A in your equation in CTQ 16? Include units.

18. As a team, devise a definition for equilibrium (as it applies to chemical reactions). Write the definition in a complete sentence or two.

19. What is one strength your team exhibited while working on this activity? Why is this a strength?

Exercises:

1. In this activity, you calculated a rate constant for the *forward* reaction (k_A) of _____ (include units).

2. In this activity, you calculated a rate constant for the *reverse* reaction (k_B) of _____ (include units).

3. Use the rate constants and the number of A and B molecues at t=50 s from Table 1 to verify that the rates of the forward reaction (A → B) and the reverse reaction (B → A) are equal.

4. The ratio of the forward rate constant to the reverse rate constant (*i. e.,* k_A/k_B) is called the *equilibrium constant*. Calculate the equilibrium constant, K_{eq}, for the reaction in Model 2.

5. When equilibrium is reached, what fraction of the molecules in the above reaction exist as products (B)?

6. Does this equilibrium *favor* (have more of) the **products** or the **reactants** (circle one)?

7. For a reaction with $K_{eq} > 1$, does the equilibrium favor the products, the reactants, or neither?

8. For a reaction with $K_{eq} < 1$, would the equilibrium favor the products, the reactants, or neither?

9. For a reaction with $K_{eq} = 1$, would the equilibrium favor the products, the reactants, or neither?

10. Read the assigned pages in your textbook, and work the assigned problems.

Le Chatelier's Principle*
(What happens when conditions change?)

Model 1: The Law of Mass Action

The Law of Mass Action states that for a chemical system described by the balanced chemical equation:

$$aA + bB \rightleftharpoons cC + dD$$

the ratio $\dfrac{[C]^c [D]^d}{[A]^a [B]^b}$ is a constant at a given temperature,

where the brackets [ ] indicate molar concentrations, or mol/L. The ratio is called the **equilibrium constant expression**, and the numerical value of the ratio is called the **equilibrium constant, K_{eq}**. Note that a, b, c, and d are the stoichiometric coefficients of the balanced chemical equation.

By convention, equilibrium constant values are given without units.

For example:

The reaction: $2 SO_2(g) + O_2(g) \rightleftharpoons 2 SO_3(g)$

The equilibrium expression: $K_{eq} = \dfrac{[SO_3]^2}{[SO_2]^2 [O_2]}$

The equilibrium constant (at 25°C): $K_{eq} = 429$ (experimentally determined)

Critical Thinking Questions:

1. a. What is the equilibrium expression for the example reaction in Model 1?

   b. If $[SO_2] = 0.150$ M and $[O_2] = 0.0751$ M, what is the concentration of $[SO_3]$?

2. Consider the equilibrium expression and the value of the equilibrium constant for the example reaction. When a mixture of SO_2, O_2, and SO_3 reaches equilibrium, does the product or do the reactants have the largest amount? (Hint: Think about the fraction. Is it larger in the numerator or the denominator?) Discuss as a team, and explain your reasoning.

3. Write the equilibrium constant expression for the reaction:

   $$2 HI \rightleftharpoons H_2 + I_2$$

---

* Adapted from ChemActivity 39, Moog, R.S. and Farrell, J.J. *Chemistry: A Guided Inquiry*, 5th ed., Wiley, 2011, pp. 226.

4. When the reaction in CTQ 3 reaches equilibrium, the following concentrations are measured: [HI] = 0.60 M; [H_2] = 0.20 M; [I_2] = 0.55 M. What is the value of K_{eq} for this reaction?

5. Would the K_{eq} calculated in CTQ 4 favor (have more of) the products or reactant? Does your team agree?

Model 2: Changes to Equilibrium

When changes occur to a system once it has reached equilibrium, the system must adjust to accommodate the change.

For the reaction: N_2 + $3 H_2$ $\rightleftharpoons$ $2 NH_3$

Initial equilibrium concentrations	5.0	7.0		5.0
Change to system		3.0 added		
Adjusted equilibrium	4.4	8.3		6.1

*All above values are molar (M).

Critical Thinking Questions:

6. What is the K_{eq} for the reaction under the initial conditions?

7. How much [H_2] is in the reaction immediately after 3.0 M is added?

8. a. To reach the new equilibrium, is H_2 used or formed? _____

   b. Which reaction (forward or reverse) will temporarily dominate to achieve the new equilibrium? Does your team agree?

9. What is the new K_{eq}?

10. As a team, predict which reaction would (temporarily) dominate if 3.0 M of H_2 had been removed?

11. Heat may be treated as a reactant or product to determine changes to equilibrium. If heat is added to the following reaction, will the forward or reverse reaction (temporarily) dominate?

$$2 SO_2(g) + O_2 \rightleftharpoons 2 SO_3 + heat$$

Information:

LeChatelier's principle says that if a system at equilibrium is changed or altered in some manner, the system will shift to accommodate the change and form a new equilibrium. When reactant or product is added, the equilibrium will shift *away* from the increase in order to use up some of the excess. When some of a reactant or product is removed, the equilibrium shifts *toward* that species to make more of it. For endothermic or exothermic reactions, heat can be treated as a reactant or product.

12. For the reaction: $N_2(g) + O_2(g) \rightleftharpoons 2\ NO(g)$

    Which direction (toward products or toward reactants) will the equilibrium shift to accomodate each of the following changes?

    a. Removal of NO

    b. Addition of O_2

    c. Removal of N_2

13. List any questions remaining with your team about equilibrium reactions.

14. Describe one area in which your team can work on improving for next time.

Exercises:

1. Write the equilibrium expression for each of the following reactions:

    a. $2\ SO_2(g) + O_2(g) \rightleftharpoons 2\ SO_3(g)$

    b. $H_2O(g) + CH_4(g) \rightleftharpoons CO(g) + 3\ H_2(g)$

2. If the $K_{eq} = 798$ at 25°C for the reaction: $2\ SO_2(g) + O_2(g) \rightleftharpoons 2\ SO_3(g)$

    calculate the equilibrium concentration of O_2 given that the concentrations of the other chemicals are: $[SO_2] = 4.20$ M; $[SO_3] = 11.0$ M

3. Consider the reaction: $H_2 + Cl_2 \rightleftharpoons 2\ HCl + heat$
    What effect does each of the following changes have on the equilibrium?

    a. add H_2

    b. add heat

    c. remove HCl

    d. remove Cl_2

4. Read the assigned pages in your textbook, and work the assigned problems.

Changes of State
(What does heat have to do with chemistry?)

Model 1: Heating curve of a pure substance

Heat energy is measured in calories (English system) or Joules (metric system).

$$1 \text{ cal} = 4.184 \text{ joules (exact)}$$

The graph (heating curve) below shows the temperature of 19 g of a pure substance that is heated by a constant source supplying 500.0 calories per minute.

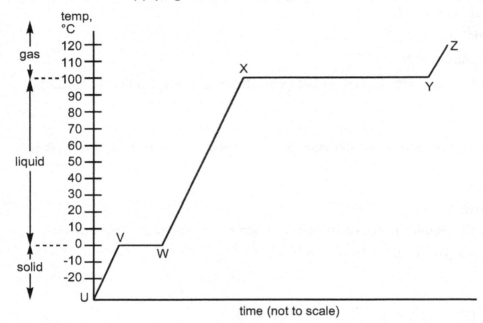

Regions: UV = 0.76 min, VW = 3.04 min, WX = 3.8 min, XY = 20.5 min, YZ = 0.63 min

Critical Thinking Questions:

1. At the coldest temperature (point U), what is the physical state of the pure substance?

    Circle one: **solid   liquid   gas**.

2. Identify the line segment on the graph (UV, VW, *etc.*) that corresponds to the pure substance:

    a. being warmed while remaining a solid _____

    b. being warmed while remaining a liquid _____

    c. being warmed while remaining a gas _____

    d. changing from a solid to a liquid _____

    e. changing from a liquid to a gas _____

3. What is the boiling point of the substance? _____

4. What is the melting point of the substance? _____

5. Why is the temperature not changing from point V to W, even though heat is being added? (Where is the heat energy going?) Does your team agree?

6. Work with your team to determine why the temperature is not changing from point X to point Y.

7. How many calories were needed to melt the solid at 0°C?

8. From your answer to CTQ 7, calculate the calories used per gram (cal/g) of this substance in order to melt it. Include units. This quantity is known as the **heat of fusion** of the substance.

9. How many calories were needed to change the liquid at 100°C to a gas?

10. From your answer to CTQ 9, calculate the calories used per gram (cal/g) of this substance to vaporize it. Include units. This quantity is known as the **heat of vaporization** of the substance. Does your team agree?

11. Work with your team to devise an explanation for why line segment XY is so much longer than line segment VW.

12. How many calories were needed to warm the liquid from point W to point X?

13. How many degrees Celsius did the temperature **change** from point W to point X?

14. From your answers to CTQs 12 and 13, calculate the calories used per gram of this substance per degree Celsius ($\frac{cal}{g\,°C}$) to change its temperature. Include units. This quantity is known as the **specific heat capacity** (or "specific heat") of the substance.

15. At what point on the curve would the molecules have the highest kinetic energy? _____

16. What do you think is the identity of the substance being heated? _____

17. Complete the definitions of the terms in **boldface** in CTQs 8, 10, and 14.

    a.   The heat of fusion of a substance is

    b.   The heat of vaporization of a substance is

    c.   The specific heat capacity of a substance is

Model 2: Specific heat capacity (or "specific heat")

The heat energy in calories (q) needed to raise the temperature of a substance may be calculated using the following equation:

$$q \text{ (heat in calories)} = s \left(\text{specific heat, } \frac{cal}{g°C}\right) \times (\text{mass in grams}) \times (\Delta T \text{ in °C})$$

or simply:

$$q = s \cdot m \cdot \Delta T$$

18. How much heat energy is required to completely change 25 grams of **liquid water** at 0°C to water at body temperature (37°C)?

19. Consider how much heat energy is required to completely change 25 grams of **ice** at 0°C to **water** at 0°C.

    a.   Circle the constant that is needed to perform this calculation:

        heat of fusion   **heat of vaporization**   **specific heat**

    b.   Perform the calculation.

20. How much total heat energy is required to completely change 25 grams of ice at 0°C to water AND also raise its temperature to body temperature (37°C)?

21. A hot iron skillet (178°C) weighing 1.51 kg is sitting on a stove. How much heat energy (in joules) must be removed to cool the skillet to room temperature, 21°C? The specific heat of iron is 0.450 J/(g·°C). Does your team agree?

22. Where does the heat energy in CTQ 21 go? _____

23. What was your plan for improving performance today? Explain why was your plan was or was not successful.

24. What are the major concepts of this ChemActivity?

Exercises:

1. Dietary calories reported on food products are actually kilocalories, written as "kcal" or "Cal" (with a capital "C"). Convert your answer to CTQ 20 into kilocalories. Would eating ice be a good way to lose weight? Explain why or why not.

2. It takes 146 J of heat energy to raise the temperature of 125.6 g of mercury from 20.0°C to 28.3°C. Calculate the specific heat capacity of mercury.

3. When 23.6 g of calcium chloride were dissolved in 300 mL of water in a calorimeter, the temperature of the water rose from 25.0°C to 38.7°C. What is the heat energy change in kcal for this process? [The specific heat of H_2O = 1.00 cal/g °C.]

4. In Exercise 3 you calculated the heat energy produced by dissolving 23.6 g of calcium chloride in water. How many **moles** of calcium chloride is this? How much heat energy is produced **per mole** of calcium chloride that dissolves in water? Report your answer in **kilocalories/mole** (kcal/mol).

5. Read the assigned pages in your text, and work the assigned problems.

Rates and Energies of Reactions
(What determines how fast a chemical reaction proceeds?)

Model 1: Enthalpy of endothermic and exothermic processes

The chemicals in a reaction **system** can either absorb heat energy from the **surroundings** or release heat energy to the surroundings. A reaction in which heat energy moves from the system to the surroundings (and feels warm) is said to be **exothermic**. A reaction in which heat energy moves from the surroundings to the system (and feels cool) is said to be **endothermic**.

The breaking of bonds requires energy from molecular collisions to separate bonded atoms (called **activation energy**, or E_{act}), and the formation of bonds releases energy to the surroundings.

Because in most chemical reactions bonds are both broken and formed, the difference in energy between the reactants and products determines whether heat is released or absorbed. This difference is called the **change in enthalpy**, or ΔH. In general, exothermic reactions favor the formation of products, and we say that they are "favorable".

a) an exothermic reaction

$$2\,H_2(g) + O_2(g) \longrightarrow 2\,H_2O(g)$$

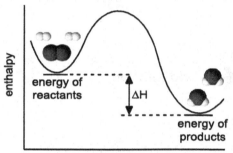

b) an endothermic reaction

$$2\,H_2O(g) \longrightarrow 2\,H_2(g) + O_2(g)$$

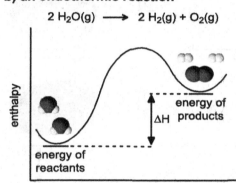

	ΔH (kcal)
Reaction (a): $2\,H_2(g) + O_2(g) \rightarrow 2\,H_2O\,(g) + heat$	− 116
Reaction (b): $2\,H_2O\,(g) + heat \rightarrow 2\,H_2(g) + O_2(g)$	+ 116

Critical Thinking Questions:

Manager: Ask a different team member to propose the first answer for each part of CTQs 1–5.

1. Compare the equations for Reactions (a) and (b) in Model 1. How is Reaction (b) related to Reaction (a)?

2. How does the difference in Reactions (a) and (b) affect the sign of ΔH?

3. For an **exothermic** reaction:
   a. Is the energy of the products **higher** or **lower** than reactants? (circle one)
   b. Is heat **released** or **absorbed** ?
   c. Is the sign for ΔH **positive** or **negative** ?

4. For an **endothermic** reaction:
   a. Is the energy of the products **higher** or **lower** than reactants? (circle one)
   b. Is heat **released** or **absorbed**?
   c. Is the sign for ΔH **positive** or **negative** ?

5. a. Since reactions are favorable when they release energy and form lower energy products, which reaction — (a) or (b) — is favorable? _____
   b. What is the sign (+ or –) for the energy (enthalpy) value of a favorable reaction? ____

Manager: Check to ensure that all team members agree on the answers to CTQs 3–5.

Model 2: Molecular collisions that may result in a chemical reaction

The hollow arrows indicate the direction and speed of motion of the atoms or molecules.

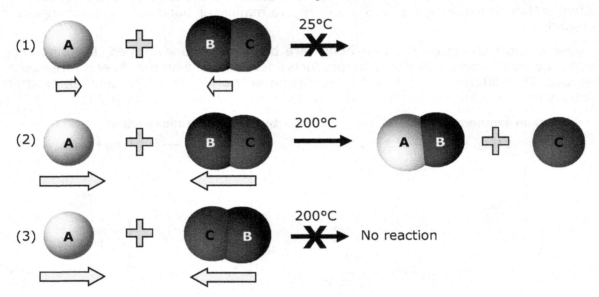

Critical Thinking Questions:

6. When the temperature is increased:
   a. Do atoms and molecules move **faster** or **slower** ? (circle one)
   b. Do atoms and molecules have **more** or **less** kinetic energy? (circle one)

7. In what way(s) are collisions between molecules at 200°C different from the collisions at room temperature (25°C)?

8. Compare reactions (1) and (2) in Model 2. In what way(s) are they different?

9. Compare reactions (2) and (3) in Model 2. In what way(s) are they different?

10. Discuss with your team, and propose at least two conditions that must be satisfied for a molecular collision to result in a chemical reaction.

Model 3: Energy diagrams for three model reactions

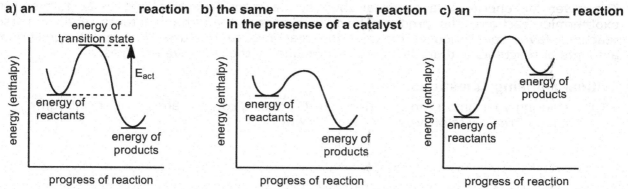

a) an _____ reaction   b) the same _____ reaction   c) an _____ reaction
                                in the presense of a catalyst

A *catalyst* increases the rate of a chemical reaction without being consumed in the reaction.

The *transition state* is the particular arrangement of atoms in the reactants at the maximum energy level.

Critical Thinking Questions:

Manager: Be sure to discuss CTQs 11–17 to ensure that all team members agree before moving on.

11. Without looking at Model 1 if possible, add the words **exothermic** and **endothermic** to the appropriate blanks in Model 3, parts (a) and (c).

12. Label the transition state energy in Model 3, parts (b) and (c).

13. Draw vertical arrows onto Models 3 (b) and (c) that represent the magnitude of the activation energy for conversion of the reactants to the products.

14. a. What do you think happens to reactant molecules that collide with **less** energy than the energy of the transition state?

   b. Which reaction in Model 2 (on the previous page) illustrates this? _____

15. In light of Models 2 and 3, why do reactions proceed faster at higher temperatures? The best answers will include the terms *molecules* and *activation energy*.

16. Examine diagram (b) in Model 3. As a team, propose an answer to complete the following sentence:

   A catalyst increases the rate of a chemical reaction by

17. Considering Model 3, does a catalyst affect the energies of the reactants? _____ Does a catalyst affect the energies of the products? _____ Indicate how you can tell this from the model.

Manager: Check that all team members agree on the answers to CTQs 11–17.

Information:

The lower the energy of a chemical species, the more favorable it is to be formed. In exothermic reactions, the products have less energy than the reactants, and so these reactions favor conversion of most of the reactants to products. The exact equilibrium amounts of reactants and products are determined by their relative energies.

Critical Thinking Questions:

18. Considering your answer to questions 16–17, does a catalyst affect the equilibrium amounts of reactants or products? Why or why not?

19. Consider Reaction (2) in Model 2. What do you think would happen to the reaction rate if more A atoms were added to the mixture, causing the concentration of A to increase? Discuss with your team and write the consensus answer.

20. As a team, propose a list of three ways to speed up a chemical reaction.

21. **Reflector (Process Analyst):** How effective was the Manager at performing his or her role today—keeping the team focused and on task, ensuring that all team members were heard, and managing timing? Write your answer in a complete sentence or two.

Exercises:

1. In a chemical equilibrium, what is it that is equal?

2. The state of equilibrium is often called a dynamic equilibrium. Explain the meaning of the term *dynamic* in this context.

3. Most exothermic reactions are considered *spontaneous*, meaning that the products are more stable (have lower energy) than the reactants. However, *spontaneous* does not necessarily equate to *fast*. Explain why a reaction that is considered spontaneous may nevertheless not show any observable reaction.

4. Sketch a plot of energy (y axis) versus reaction progress (x axis) for a typical *exothermic* reaction. Then, draw another line onto the *same* plot which shows how the curve changes when the reaction occurs in the presence of a catalyst. Label each line. Explain how a catalyst increases the rate of a chemical reaction in terms of the meaning of your plot.

5. Consider the reaction in Exercise 4:
   a. Would this reaction *favor* formation of mostly **products** or mostly **reactants**?
   b. Would the equilibrium constant, K_{eq}, for this reaction be **less than**, **equal to**, or **greater than** 1? Circle one, and explain your choice.

   c. When this reaction has reached equilibrium, would the forward rate be **less than**, **equal to**, or **greater than** the reverse rate? Circle one, and explain your choice.

6. The burning of a piece of paper in air (to produce CO_2 and H_2O vapor) is a spontaneous exothermic reaction.
   a. What would be necessary to start this reaction?

   b. After the reaction starts, what provides the activation energy for the reaction to continue?

7. Read the assigned pages in your textbook and work the assigned problems.

CA25

Gases
(Do all gases behave the same?)

Model 1: Representation of the molecular view of water in a closed container at 20°C (293 K).

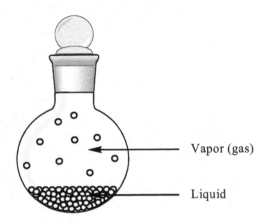

Vapor (gas)

Liquid

Critical Thinking Question:

1. Considering Model 1, work with your team to compose a short list of the main differences between a gas and a liquid at the same temperature.

Model 2: Representation of some gases and their properties under different conditions.

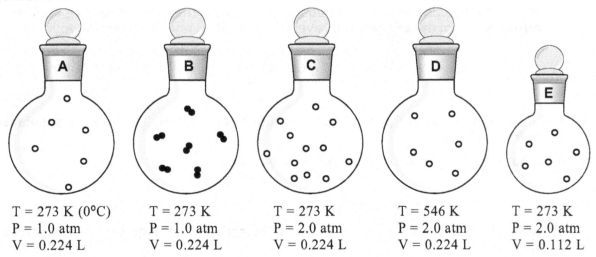

A	B	C	D	E
T = 273 K (0°C)	T = 273 K	T = 273 K	T = 546 K	T = 273 K
P = 1.0 atm	P = 1.0 atm	P = 2.0 atm	P = 2.0 atm	P = 2.0 atm
V = 0.224 L	V = 0.224 L	V = 0.224 L	V = 0.224 L	V = 0.112 L

Key: T = temperature; P = Pressure; V = volume; ● = 1.0×10^{21} molecules N_2; ○ = 1.0×10^{21} atoms He

The standard air pressure at sea level (about 15 lb/in^2) is <u>1 atmosphere</u> (atm)

1 atm = 760 mm Hg = 760 torr

Critical Thinking Questions:

2. Compare containers A and B in Model 2. How does the pressure (P) change when the identity of the gas in the container is changed at constant volume (circle one)?

   P is doubled   **P is halved**   **P remains constant**

3. Compare containers A and C in Model 2. How does the pressure of a gas change when the number of **atoms** in the container is doubled at constant volume?

   P is doubled   **P is halved**   **P remains constant**

4. The number of atoms and the number of moles are proportional. How does the pressure of a gas change when the number of **moles** of gas in the container is doubled at constant volume? Does your team agree?

   P is doubled   **P is halved**   **P remains constant**

5. Considering Model 2, how does the pressure of a gas change when the temperature is doubled at constant volume (containers A and D)?

6. Considering Model 2, how does the pressure of a gas change when the volume of the container is doubled at constant temperature (containers A and E)?

Manager: *Identify a different team member to explain CTQs 7 – 9.*

7. The symbol $\propto$ means "is proportional to," and the symbol "n" means "number of **moles** of gas." Based on your answer to CTQ 4, circle the correct expression.

$$P \propto n \qquad\qquad P \propto \frac{1}{n}$$

8. Based on your answer to CTQ 5, circle the correct expression.

$$P \propto T \qquad\qquad P \propto \frac{1}{T}$$

9. Based on your answer to CTQ 6, circle the correct expression.

$$P \propto V \qquad\qquad P \propto \frac{1}{V}$$

10. How many atoms of He are there in container A in Model 2? _____

11. Using the number of atoms in container A in Model 2, calculate the number of moles (n). Recall that Avogadro's number is 6.022×10^{23}.

Model 3: The Ideal Gas Law

Rearranging the expressions from CTQs 7-9 gives an equation that relates all four variables: pressure (P), volume (V), temperature (T) and number of moles (n).

$$PV = nRT$$
where R is a constant called the *universal gas constant*

Critical Thinking Questions:

12. Divide both sides of the Ideal Gas Law by V, and write the result. Is the Ideal Gas Law consistent with your answers to CTQs 7-9?

13. Divide both sides of the Ideal Gas Law by n and T to solve it for R. Then use the values of P, T, n, and V from container A in Model 2 to calculate R. Give the correct units.

14. Repeat CTQ 13 using the values of P, T, n, and V from container C, D, or E in Model 2 (your choice). Again, keep track of the units.

15. Compare your answers for CTQs 13 and 14. Are they the same? Discuss with your team, and explain why or why not.

16. Using the Ideal Gas Law, calculate the volume of 20.5 g of NH_3 at 0.658 atm and 25°C. Recall the conversion for temperature K = °C + 273. (Remember your units for R). Compare your answer with your team and/or other teams.

17. As a team, reflect on your work so far. What is one way in which you are working well together as a team, and what is one way you can improve? Outline a strategy for improvement for the remainder of the activity.

Model 4: Mixtures of gases

In a mixture of gases, the total pressure, P_T, is the sum of the pressures of the individual gases. The individual pressures are called **partial pressures**.

$$P_T = P_1 + P_2 + P_3 + P_4 + \ldots \qquad (1)$$

The pressure of each gas in the mixture obeys the ideal gas law:

$$P_i = n_i \frac{RT}{V} \qquad (2)$$

Critical Thinking Questions:

18. Is equation (2) from Model 4 in agreement with the ideal gas law? Explain.

19. A student makes the following statement: "The ratio of the partial pressures of two gases in a mixture is the same as the ratio of the number of moles of the two gases." Is the student correct? Explain your answer.

Model 5: The Combined Gas Law

Rearranging the Ideal Gas Law gives an equation that allows one to calculate the change in one variable caused by changes in the other two.

$$\frac{P_1 V_1}{T_1} = \frac{P_2 V_2}{T_2}$$

The subscripts denote a gas under one set of conditions (1) changed to another set (2).

Critical Thinking Questions:

20. If T is held constant in Model 5, what is the resulting equation?

21. What is the effect on the volume of a gas if you simultaneously double its pressure and double its Kelvin temperature? Discuss your answer with your team.

22. A sample of nitrogen gas has a volume of 3.5 L at 1.5 atm. If the temperature remains the same, what volume would the gas have if the pressure dropped to 0.30 atm? Ensure that all team members agree.

23. What are the major concepts in this activity?

24. List one or two insights or strategies for solving gas law problems that your team discovered today.

25. Did everyone in your team understand the material in the activity today? If so, explain how your team ensured that everyone understood. If not, identify what your team needs to do to assure that everyone in the team understands the material in the next session.

Exercises:

1. Calculate the volume of 359 g of ethane (C_2H_6) at 0.658 atm and 75°C. (Watch the units!)

2. A 10.00-L SCUBA tank holds 18.0 moles of O_2 and 12.0 moles of He at 289 K. What is the partial pressure of O_2? What is the partial pressure of He? What is the total pressure?

3. The 10.00 L SCUBA tank in Exercise 2 is filled to a pressure of 120 atm at 20°C. What is the pressure of air in the tank when the temperature drops to −10°?

4. Read the assigned pages in your text, and work the assigned problems.

Intermolecular Forces[*]
(Why do liquids stick together?)

Model 1: Water at room temperature (A) and after heating (B)

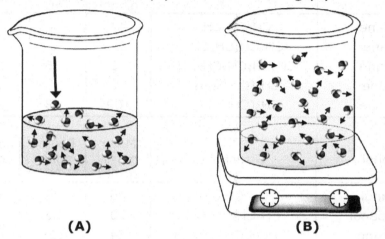

(A)   **(B)**

Critical Thinking Questions:

1. In Model 1A, is the water mainly in a solid, liquid, or gaseous state? _____

2. What do the small arrows represent in Model 1?

3. Note the water molecule in Model 1A indicated by the large arrow. What is different about this molecule compared with the others in the container?

4. Consult with your team, and propose an explanation for how the molecule in CTQ 3 obtained its location.

5. When water evaporates, are any O−H bonds within a water molecule broken?

6. How does the process of evaporation change when heat is applied (Model 1B)?

7. On average, are the water molecules closer together in $H_2O(l)$ or $H_2O(g)$? _____

8. On average, would you expect the attractive forces between the molecules (intermolecular forces) to be stronger in $H_2O(l)$ or $H_2O(g)$? Does your team agree? Explain your choice.

---

[*]Adapted from ChemActivity 27, Moog, R.S.; Farrell, J.J. *Chemistry: A Guided Inquiry*, 5th ed., Wiley, 2011, pp. 157-163.

Model 2: Boiling points and water solubility of selected compounds[†]

To a large extent, the boiling point of a liquid is determined by the strength of the *intermolecular attractions* in the liquid. These attractions are determined by the size and structure of the individual molecules.

Alkane	Structure	Molar mass, g/mol	Boiling point, °C	Water solubility, grams per 100 mL H_2O
propane	$CH_3CH_2CH_3$	44	−42	0.007
butane	$CH_3CH_2CH_2CH_3$	58	0	0.006
pentane	$CH_3CH_2CH_2CH_2CH_3$	72	36	0.04
hexane	$CH_3(CH_2)_4CH_3$	86	69	0.001
heptane	$CH_3(CH_2)_5CH_3$	100	98	0.01

Alcohol	Structure	Molar mass, g/mol	Boiling point, °C	Water solubility, grams per 100 mL H_2O
ethanol	CH_3CH_2OH	46	78	∞
1-propanol	$CH_3CH_2CH_2OH$	60	82	∞
1-butanol	$CH_3CH_2CH_2CH_2OH$	74	118	6.3
1-pentanol	$CH_3(CH_2)_4OH$	88	137	2.7
1-hexanol	$CH_3(CH_2)_5OH$	102	157	0.6

Amine	Structure	Molar mass, g/mol	Boiling point, °C	Water solubility, grams per 100 mL H_2O
propyl amine	$CH_3CH_2CH_2NH_2$	59	48	∞
butyl amine	$CH_3CH_2CH_2CH_2NH_2$	73	77	∞
hexyl amine	$CH_3(CH_2)_5NH_2$	101	131	1.2
octyl amine	$CH_3(CH_2)_7NH_2$	129	180	0.02

Ether	Structure	Molar mass, g/mol	Boiling point, °C	Water solubility, grams per 100 mL H_2O
dimethyl ether	CH_3OCH_3	46	−23	∞
diethyl ether	$CH_3CH_2OCH_2CH_3$	74	35	6.9
dipropyl ether	$CH_3(CH_2)_2O(CH_2)_2CH_3$	102	89	0.25

Note: ∞ means infinitely soluble (*i. e.,* the liquids are miscible)

Critical Thinking Questions

9. Recall that the electronegativity of C and H are roughly the same, but that N and O have significantly higher electronegativities. For each type of compound (alkane, alcohol, amine, ether) predict whether or not the compound is expected to be polar or nonpolar. Does your team agree?

[†] Sources: CRC Handbook of Chemistry and Physics, 47th ed., 1967; ChemFinder.com; IPCS INCHEM, www.inchem.org; Korea thermophysical properties Data Bank, www.cheric.org/kdb/ [accessed Jan 2012]

10. For a given type of compound (alkane, alcohol, amine or ether):
   a. How does the boiling point change as the molar mass of the compound increases?

   b. Do the intermolecular forces (attractions between) molecules **increase** or **decrease** as the molar mass increases? Discuss with your team and explain your choice.

11. Work with your team to choose an alkane, an alcohol, an amine and an ether with roughly the same molar mass (within 5 g/mol). Rank these compounds in terms of relative boiling points.

12. Repeat CTQ 11 with another set of four compounds.

13. Does polarity in molecules tend to **increase** or **decrease** the strength of intermolecular interactions? Write a team consensus explanation of your reasoning.

14. Based on relative boiling points, write the numbers from 1 to 4 under the functional group names below, with the number 1 being the group with the most intermolecular attractions, and 4 being the least.

    alkane   alcohol   ether   amine

15. For each functional group below, circle each type of bond contained in the molecule. You may refer to Model 2.

functional group	Type of bond			
alkane	C–H	C–O	O–H	N–H
alcohol	C–H	C–O	O–H	N–H
ether	C–H	C–O	O–H	N–H
amine	C–H	C–O	O–H	N–H

16. Based on your answers to CTQs 14 and 15, which two types of bonds are present only in molecules containing the two functional groups with the strongest intermolecular attractions?

Model 3: Intermolecular forces are weaker than covalent bonds

The types of bonds you identified in CTQ 16 can exhibit an intermolecular force known as **hydrogen bonding**. A "hydrogen bond" is simply a particularly strong attraction between the bonded hydrogen atom and a lone pair on another atom.

The intermolecular forces that attract molecules to each other are much weaker than the bonds that hold molecules together. For example, 463 kJ/mol are required to break one mole of O−H bonds in H_2O molecules, but only 44 kJ/mol are needed to separate one mole of water molecules in liquid water.

Types of intermolecular forces:

Hydrogen bonding: An interaction between a hydrogen atom attached to an electronegative atom of **fluorine, oxygen, or nitrogen** on one molecule and **a lone pair of electrons** on a fluorine, oxygen, or nitrogen on another molecule. This is the strongest intermolecular force.

Dipole-dipole forces: Often polar covalent bonds create polar molecules which have a positive end and a negative end. The positively polarized ends of molecules will attract the negatively polarized ends of other molecules.

London dispersion forces: At any given instant, electrons may be nearer to one end of a molecule or the other. When they are nearer to one end, they repel electrons in nearby molecules, causing the ends of the molecules nearby to have a partial positive charge, and creating a momentary attractive force. This is the weakest force and is present in all molecules. In large molecules, this force can be quite significant.

Critical Thinking Questions:

17. What is the difference between intramolecular bonds and intermolecular forces?

18. Rank these intermolecular forces in terms of their typical relative strengths: dipole-dipole, hydrogen bonding, London forces. Compare with your team.

19. In terms of intermolecular forces, write a consensus explanation of the general trend that you described in CTQ 13.

20. Suppose that we consider anything over about <u>1 gram per 100 mL water</u> to be "soluble."

    a. What is the heaviest alcohol in Model 2 that is water soluble? _____

    b. How many carbons are in this alcohol? ____

    c. What is the heaviest amine in Model 2 that is water soluble? _____

    d. How many carbons are in this amine? ____

21. Considering your answers to CTQ 20, the presence of <u>one polar functional group</u> is sufficient to dissolve a molecule containing about how many nonpolar carbons? Circle one of the following choices:

1−2   3−4   5−6   7−8

Model 4: Water solubility of selected ketones

Ketone	Structure	Water solubility, g per 100 mL H₂O	Ketone	Structure	Water solubility, g per 100 mL H₂O
acetone	$H_3C-\overset{O}{\overset{\|\|}{C}}-CH_3$	∞	2-hexanone	$H_3C-\overset{O}{\overset{\|\|}{C}}-(CH_2)_3CH_3$	1.4
2-butanone	$H_3C-\overset{O}{\overset{\|\|}{C}}-CH_2CH_3$	25.6	2-heptanone	$H_3C-\overset{O}{\overset{\|\|}{C}}-(CH_2)_4CH_3$	0.4
2-pentanone	$H_3C-\overset{O}{\overset{\|\|}{C}}-CH_2CH_2CH_3$	4.3			

Critical Thinking Questions:

22. Consider the water solubilities of the ketones shown in Table 2. Are they consistent with your answer to CTQ 21? Write a sentence that generalizes how many nonpolar carbons can be dissolved by one polar functional group.

23. Consider the examples of molecules below.

$H_3C-\overset{H_2}{\overset{\|}{C}}-CH_3$ $H_3C-\overset{H_2}{\overset{\|}{C}}-\overset{H_2}{\overset{\|}{C}}-\overset{..}{O}H$ $H_3C-\overset{..}{\underset{..}{O}}-CH_3$ $H_3C-\overset{H_2}{\overset{\|}{C}}-\overset{..}{N}H_2$ $H_3C-\overset{\overset{..}{\overset{O}{\|\|}}}{C}-CH_3$

    alkane   alcohol   ether   amine   ketone

   a. Which of the molecules are polar?

   b. Which of the molecules can form hydrogen bonds to other identical molecules?

   c. Which of the molecules can form hydrogen bonds with water?

24. What is one area your team can improve on for the next class period? How do you plan to accomplish this?

Exercises :

1. Define intermolecular forces.

2. Describe what happens to intermolecular forces when a liquid is heated.

3. What type(s) of intermolecular force(s) are present in alkanes? What is the strongest intermolecular force present?

4. What type(s) of intermolecular force(s) are present in alcohols? What is the strongest intermolecular force present?

5. What type(s) of intermolecular force(s) are present in amines? What is the strongest intermolecular force present?

6. What type(s) of intermolecular force(s) are present in ethers? What is the strongest intermolecular force present?

7. What type(s) of intermolecular force(s) are present in ketones? What is the strongest intermolecular force present?

8. The ketone 2-hexanone and the alcohol 1-hexanol have similar molar masses. The boiling point of the ketone is about 20°C less than the alcohol. However, the water solubility of the ketone is slightly higher. Explain this result in terms of the intermolecular forces involved.

9. Both cis-1,2-dichloroethylene and trans-1,2-dichloroethylene have the same molecular formula: $C_2H_2Cl_2$. However the cis compound is polar, while the trans compound is not. One of these species has a boiling point of 60.3°C and the other has a boiling point of 47.5°C. Which compound has which boiling point? Explain your reasoning.

10. Fluoromethane, CH_3F, and methanol, CH_3OH, have approximately the same molecular weight. However the boiling point of CH_3OH is 65.15°C, whereas the boiling point of CH_3F is almost 100 degrees lower, -78.4°C. Explain this result.

11. A hydrogen bond is usually depicted by a dashed or dotted line. Circle the picture that correctly represents a hydrogen bonding interaction.

12. Describe what is wrong with each of the other three pictures in Exercise 11.

13. Draw a representation of one water molecule participating in a hydrogen bond with another water molecule. Then place the symbols $\delta+$ and $\delta-$ near each atom to indicate the polarity of the bonds.

14. Without looking up any information in a table, identify the molecule in each group that would have the highest boiling point, and explain your answer.

    a. CH_3CH_2OH $CH_3CH_2CH_2OH$ $HOCH_2CH_2OH$

                    OH   OH   OH

    b. $H_3C-CHCH_3$ $H_3C-CHCH_2CH_2CH_3$ $CH_3CH_2-CHCH_3$

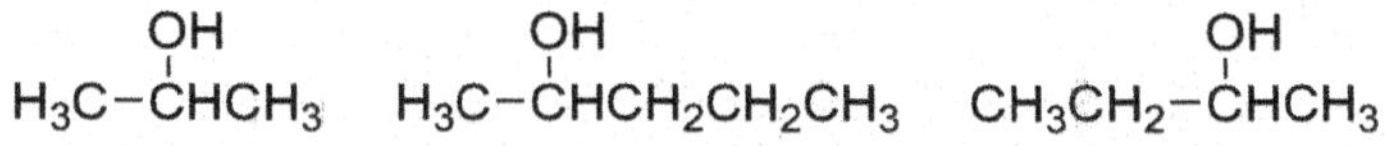

    c. $CH_3CHOHCH_2CH_3$ $CH_3CH_2CH_2CH_2CH_3$

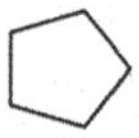

    d.

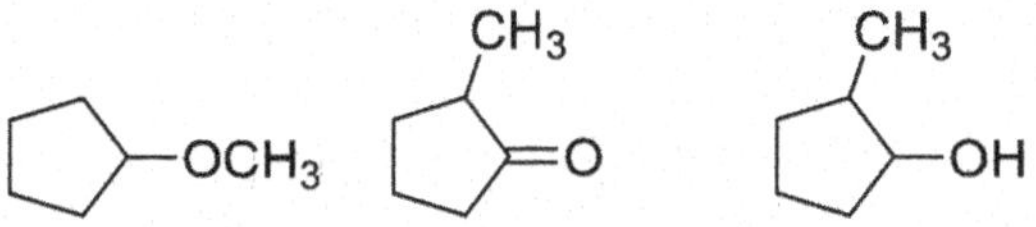

15. Benzoate ion is very soluble in water, but benzoic acid is not. Based on this information which species do you think has a larger polarity—benzoic acid or benzoate? Explain.

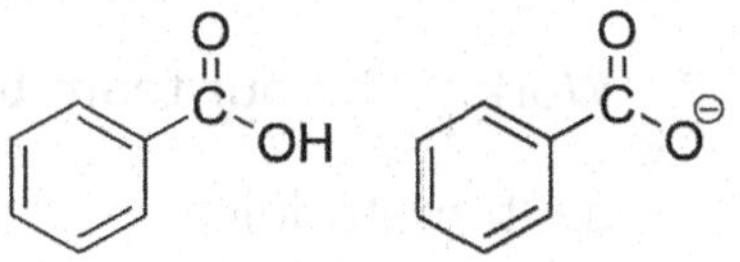

benzoic acid   benzoate

16. Read the assigned pages in your text, and work the assigned problems.

CA27

Solutions and Concentration
(When is it dissolved, and when is it suspended?)

Information:

When two substances are mixed together and the mixture is homogeneous, we call the mixture a **solution**. The component of a solution in the greater amount is the **solvent**, and the component in the lesser amount is the **solute**. The most common solvent on earth is water. Solutes that dissolve in water may be solids (*e. g.*, salt or sugar), liquids (*e. g.*, alcohol) or gases (*e. g.*, ammonia or oxygen).

Each solute has a particular **solubility** in water—often reported as the maximum number of grams of the solute that will dissolve in 100 grams of water.

If the solute particles are large enough to be seen (they scatter light, making the mixture *cloudy*) but not to settle out, the mixture is called a **colloid**. If the particles are larger still and can settle out over time, the mixture is called a **suspension**.

Critical Thinking Questions:

1. Based on the information above (or perhaps a demonstration by your instructor), work with your team to describe the differences between a solution, colloid, and suspension.

2. In each of the following mixtures, name the solvent and at least one solute.

Mixture	Solvent	Solute(s)
fog		
apple juice		
cola		
NaCl(aq)		

3. Work with your team to label each of the following as a solution, colloid, or suspension.

   a. tomato juice

   b. fog

   c. apple juice

   d. Italian salad dressing

   e. tea

   f. muddy water

   g. homogenized milk

   h. cola

Model 1: Molarity

The measure of the amount of solute dissolved in a specified amount of solution is called the **concentration** of the solution. The most common measure of concentration is **molarity**.

Molarity (M) – the moles of **solute** per liter of **solution**: $M = \dfrac{moles\ of\ solute}{L\ of\ solution} = \dfrac{mol}{L}$

For example, battery acid is approximately 6 M H_2SO_4. This means that there are 6 moles of sulfuric acid in every liter of solution. We can write two conversion factors:

$$\frac{6\ moles\ H_2SO_4}{1\ L} \quad and \quad \frac{1\ L}{6\ moles\ H_2SO_4}$$

Note how we assume—but didn't actually *write*—the word "solution" with the "1 L."

How to make 100 mL of a 2.0-molar (2.0 M) aqueous solution of NaCl

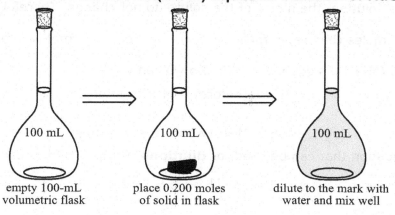

empty 100-mL
volumetric flask

place 0.200 moles
of solid in flask

dilute to the mark with
water and mix well

Note: volumetric flasks come in standard sizes, such as 25 mL, 50 mL, 100 mL, 250 mL, 500 mL, 1 L, *etc.*

Critical Thinking Questions:

4. How many moles of H_2SO_4 are found in a liter of 6 M H_2SO_4? In one liter of 12 M H_2SO_4?

5. In Model 1, why does it only require two-tenths of a mole of sodium chloride to make a 2.0-molar solution?

6. Model 1 shows 0.200 moles of NaCl in the flask; however, moles cannot be measured directly in the lab.

    a. How must be the solid be measured instead?

    b. How much solid would be measured out for the solution in Model 1?

7. An experiment calls for 50 mL of a 0.50 M aqueous solution of sodium hydrogen carbonate (sodium bicarbonate, or baking soda). Work with your team to *describe* (with amounts) the steps you would use in order to make up such a solution, such that you have none left over.

Model 2: Dilutions

When **diluting** a solution, the moles of the solute do not change. Therefore,

$$\text{moles (before)} = \text{moles (after)} \qquad or \qquad mol_1 = mol_2$$

Since M= mol/L, (M × L) is equal to moles. So we can write:

$$M \times L \text{ (before)} = M \times L \text{ (after)}$$
$$or$$
$$M_1V_1 = M_2V_2 \text{ (where V = volume)} \qquad\qquad (1)$$

This is an equation that can be used for **dilutions**. It may also be rearranged to give:

$$M_2 = \frac{M_1V_1}{V_2} \qquad\qquad (2)$$

How to make 100 mL of 0.060 M CuSO₄ from 0.60 M CuSO₄

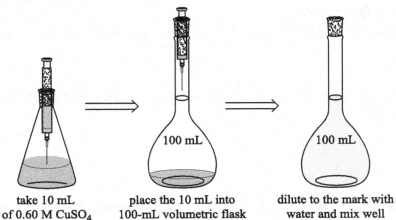

take 10 mL   place the 10 mL into   dilute to the mark with
of 0.60 M CuSO₄   100-mL volumetric flask   water and mix well

Critical Thinking Questions:

8. Consider Model 2. Verify that diluting 10 mL of 0.60 M copper(II) sulfate to 100 mL will produce a 0.060 M solution. (Identify M_1, V_1, and V_2 and calculate M_2).

9. How many mL of 0.60 M copper(II) sulfate would need to be added to the flask and diluted to a total volume of 100 mL to make an 0.1 M solution? (Hint: Which variable in equation (1) is not known?)

Information:

Electrolytes in blood are often measured in equivalents per liter (Eq/L). An **equivalent** of an ion is the amount of that ion that gives 1 mole of positive or negative charge. This gives rise to conversion factors such as:

$$1 \text{ mol Na}^+ = 1 \text{ Eq Na}^+ \qquad 1 \text{ mol Ca}^{2+} = 2 \text{ Eq Ca}^{2+} \qquad 1 \text{ mol SO}_4^{2-} = 2 \text{ Eq SO}_4^{2-}$$

Critical Thinking Questions:

10. a. How many grams are in one mole of Ca^{2+}?

    b. How many grams are in one equivalent of Ca^{2+}?

11. A patient has a blood calcium level of 4.6 mEq/L.

    a. Write the two conversion factors for converting moles of calcium into equivalents of calcium.

    b. What is the patient's blood calcium level in molarity?

12. Write any questions remaining with your team about molarity or equivalents.

Exercises:

1. If 50 mL of concentrated (18 M) sulfuric acid is diluted to a total volume of 1.0 L, what is the new concentration?

2. Iodine dissolves in various organic solvents, such as dichloromethane (CH_2Cl_2), in which it forms an orange solution. What is the molarity of I_2 when 5.00 g iodine is dissolved in enough dichloromethane to make 50.0 mL of solution?

3. A patient's blood calcium level is 9.2 mg/dL. What is this concentration in molarity? Make a unit plan first.

4. Read the assigned pages in your textbook, and work the assigned problems.

Hypotonic and Hypertonic Solutions
(What is normal saline?)

Information: Units of concentration

The *concentration* of a solution is a measure of the amount of solute dissolved in a specified amount of solution. A solution that is more *concentrated* has more solute per unit of volume than one that is more *dilute*.

We saw in ChemActivity 28 that the most common measure of concentration is molarity.

Molarity (M) – the moles of solute per liter of solution: $M = \dfrac{moles\ of\ solute}{L\ of\ solution} = \dfrac{mol}{L}$

The other common measures of concentration are not done using moles, but using either mass (weight) or volume. These are commonly reported in percent. "Percent by weight" and "percent by volume" are common terms, and may be represented as %(w:w) or %(v:v). Weights (or masses) are in grams, and volumes in milliliters. The first letter in the parenthesis represents the units of the **solute**, and the second is for the **solution**.

Since *percent* means *"per hundred,"* a 3.2%(v:v) aqueous solution of alcohol would mean 3.2 mL of alcohol are in every 100 mL of solution. We can write this as a conversion factor:

$$\frac{3.2\ \text{mL alcohol}}{100\ \text{mL}} \quad \text{and} \quad \frac{100\ \text{mL}}{3.2\ \text{mL alcohol}}$$

Note how we assume—but didn't actually *write*—the word "solution" with the "100 mL."

Critical Thinking Questions:

1. Write the two conversion factors for each of the following concentrations.

    a. 10%(v:v) acetone in water

    b. 0.90%(w:v) NaCl(aq)

    c. 5.0%(w:w) NaHCO₃(aq)

    d. 5.0%(w:v) NaHCO₃(aq)

2. The solutions in CTQ 1 (c) and (d) are often considered to be the same concentration even though the units are different. Work with your team to devise an explanation for this. (Hint: What is the density of water?)

3. Sometimes when concentrations are reported for aqueous solutions of solids, they do not say whether they are by weight or by volume. Explain why it is probably okay to assume they are %(w:v). Ensure that all team members agree.

4. Percent means parts per hundred. We can also report concentrations in parts per thousand (ppt), parts per million (ppm), or parts per billion (ppb). The EPA safe limit for lead in drinking water is 15 ppb. Write the two conversion factors for this concentration. Ensure that all team members agree.

Model 1: Movement of a solvent through a membrane

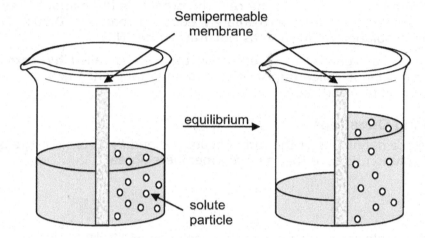

Solvent molecules are so numerous in the solutions in the Model that they are not shown individually.

Critical Thinking Questions:

5. Suppose that water is the solvent in the above model. What happens to the water on the left side of the membrane as the solution reaches equilibrium?

6. What does the semipermeable membrane allow to pass through?

7. What does the semipermeable membrane *not* allow to pass through?

8. Which side of the membrane has the higher concentration of solvent (water) at the beginning? Discuss with your team to develop a shared understanding of your team's answer.

9. As the solution reaches equilibrium (circle your choices):

   a. Water has moved *out of* an area of **higher** or **lower** *solvent* concentration.

   b. Water has moved *into* an area of **higher** or **lower** *solvent* concentration.

   c. Water has moved *into* an area of **higher** or **lower** *solute* concentration.

10. At equilibrium, water moves across the membrane in both directions; however, the rate of flow is faster to the solute side. Work with your team to propose a reason why this is the case.

Information: Hypotonic and hypertonic solutions

The process in Model 1, in which a solvent crosses a semipermeable membrane from a solution of lower solute concentration to one of higher solute concentration, is called **osmosis**.

The flow of solvent can be stopped or even reversed if pressure is applied to the solution on the side of higher solute concentration. The amount pressure required to just stop the flow of solvent without reversing the flow is called **osmotic pressure**.

Osmotic pressure can cause cells to shrink or swell. The greater the difference in solute concentration from inside to outside the cell, the greater is the osmotic pressure. Red blood cells have an osmotic pressure approximately equal to that of a 0.90% (w:v) solution of sodium chloride, a solution referred to as **normal saline** (NS).

A solution that has a higher ion concentration than NS is called **hypertonic**; one with a lower ion concentration is called **hypotonic**. It is important when giving fluids intravenously that they be isotonic to blood cells.

Critical Thinking Questions:

11. Considering the definitions of the terms *hypotonic* and *hypertonic*, write a definition for an *isotonic* solution. Ensure that all team members agree.

12. If a semipermeable membrane separates a solution of 1% NaCl and a solution of 10% NaCl, in which direction will water move?

13. Work with your team to determine what would happen to red blood cells (shrink, stay the same, swell) if they were placed in the following solutions:

    a. 10% (w:v) NaCl?

    b. pure water

14. Work with your team to determine how you would prepare an intravenous solution that would be *isotonic* with red blood cells.

15. Often the principle of osmosis is used to remove undesirable solutes such as urea from the blood of individuals with kidney problems. Discuss with your team and propose a method to prepare a solution that would enable the removal of urea without changing the blood in any other manner. (Assume your membrane will allow small molecules to pass through in addition to solvent).

16. Summarize the main points in this ChemActivity.

17. List any remaining questions about the topics in this ChemActivity.

18. List one strength of your teamwork today and why it helped the learning process.

Exercises:

1. If *percent* means *parts per hundred*, what would you call one part in 10^6?

2. If the lead level in some drinking water is 15 ppb, how many mL of the water would a person have to drink in order to ingest one gram of lead? (Unit plan!)

3. An experiment calls for 10.0 mL of a 4.00% aqueous solution of sodium tetraborate ("borax"). Describe how you would make up such a solution so that you will not have any left over.

4. Is a 1.0 M solution of NaCl hypotonic, hypertonic, or isotonic to red blood cells? (Hint: Convert to % w:v—make a unit plan!)

5. Is a 0.9 M $CaCl_2$ solution isotonic to red blood cells? Explain why or why not, without doing a calculation.

6. Indicate which side will increase in volume (A or B) for the following pairs separated by a semipermeable membrane:

A	B
a. 4%(w:v) glucose	8%(w:v) glucose
b. 2.5%(w:v) albumin	10%(w:v) albumin
c. 10%(w:v) NaCl	0.1%(w:v) NaCl

7. Read the assigned pages in your textbook, and work the assigned problems.

Colligative Properties
(How do solutes affect temperatures?)

Model 1: Boiling and freezing points of selected solutions (1 L total volume)

Solution	Boiling Point	Freezing Point
Water	100°C	0.0°C
1.0 M NaCl	101°C	−3.8°C
2.0 M NaCl	102°C	−7.6°C
1.0 M glucose	100.5°C	−1.9°C
1.0 M KBr	101°C	−3.8°C
1.0 M CaCl₂	101.5°C	−5.7°C
1.0 M Na₂S	101.5°C	−5.7°C

Critical Thinking Questions:

1. Does the boiling point of water increase or decrease with the addition of solute?

2. Does the freezing point of water increase or decrease with the addition of solute?

3. By how much does the boiling point of 1 L of water change with the addition of one mole of NaCl?

4. By how much does the boiling point of 1 L of water change with the addition of one mole of glucose?

5. As a team, discuss and predict a reason for the difference between the changes caused by 1.0 M NaCl and 1.0 M glucose. (Hint: The formula of glucose is $C_6H_{12}O_6$. What is the major difference?)

6. What difference, if any, is there in the effects of 1.0 M NaCl and 1.0 M KBr on the boiling point?

7. What difference, if any, is there in the effects of 1.0 M NaCl and 1.0 M CaCl₂ on the boiling point?

8. Discuss with your team and predict a reason for the difference between the changes caused by NaCl and CaCl₂.

9. Discuss with your team and devise a rule for how to predict the change in boiling point of an aqueous solution of any compound. When your team agrees, write your statement in a complete sentence.

10. What is the magnitude of change in the freezing point of 1 L of water when one mole of NaCl is added?

When two moles are added?

11. Discuss with your team and devise a rule for how to predict the change in freezing point of an aqueous solution of any compound. When your team agrees, write your statement in a complete sentence.

12. In cold climates, salts are spread onto highways to melt any ice that accumulates. If you choose a salt based on which one would require the least amount, would you prefer NaCl or $CaCl_2$? Does your team agree?

Information:

Colligative properties are those that depend only on the concentration of a dissolved solute (ions or molecules) and not on its chemical identity. Therefore, a mole of a covalent compound like glucose will raise the boiling point of 1 L of water by only 0.5°C, while a mole of ionic compounds like NaCl and $CaCl_2$ can raise it by 1.0°C and 1.5°C respectively.

Critical Thinking Questions:

13. What is the boiling point of 2.5 M glucose?

14. What is the boiling point of 0.50 M KBr?

15. What is(are) the most important concept(s) from today?

16. Explain how the concept of freezing point depression can help explain how highway departments treat roads during winter storms.

17. Identify one question remaining with your team about the topics in this ChemActivity.

18. Indicate one strength of your teamwork today and why it was beneficial to the learning process.

Exercises:

1. Which solution has the lower freezing point, 1.0 M KBr or 1.0 M $CaBr_2$?

2. What would be the boiling points of the solutions in Exercise 1?

3. Which solution has the higher boiling point, 0.75 M glucose or 0.6 M KCl?

4. Antifreeze in cars can help prevent the engine from overheating in warm climates. If ethylene glycol ($C_2H_6O_2$) is the main component, how many grams of ethylene glycol would need to be added to 2 L of water to raise the boiling point to 110°C?

5. Rock salt (NaCl) is added to ice to provide a low enough temperature to make homemade ice cream. What percent NaCl solution needs to be achieved to lower the temperature of the ice/water mixture to −20°C?

6. Read the assigned pages in your textbook, and work the assigned problems.

Acids and Bases
(What happens when hydrogen ions are transferred between species?)

Model 1: Some acid-base reactions

acid		base		conjugate base		conjugate acid	
CH_3COOH	+	H_2O	$\rightleftharpoons$	CH_3COO^-	+	H_3O^+	(1)
HCl	+	NH_3	$\rightleftharpoons$	Cl^-	+	NH_4^+	(2)
H_3O^+	+	HCO_3^-	$\rightleftharpoons$	H_2O	+	H_2CO_3	(3)
H_2O	+	$H_2PO_4^-$	$\rightleftharpoons$	OH^-	+	H_3PO_4	(4)

Critical Thinking Questions:

1. Examine the reactions Model 1. What happens to the species that acts as an acid (left column) as it undergoes the reaction to become the conjugate base?

2. What happens to the base as it undergoes the reaction to become the conjugate acid?

3. As a team, review your answers to CTQs 1 and 2. Complete the following sentences.

   In an acid-base reaction, the acid _____.

   In an acid-base reaction, the base _____.

4. Label the acid and the base in the reaction below, using your definitions from CTQ 3 to identify the acid and base in the reactants.

   $HNO_2 + H_2O \rightleftharpoons NO_2^- + H_3O^+$

5. An *amphoteric* substance is one that can act as either an acid or a base, depending on the situation. Review the reactions in Model 1 and identify any amphoteric substances.

6. Review reaction (1) in Model 1. What is the difference between an acid and its conjugate base?

7. Review reaction (1). What is the difference between a base and its conjugate acid?

Model 2: Conjugate acid-base pairs

According to the Brønsted-Lowry theory, a reaction of an acid and a base involves a proton (*i. e.,* hydrogen ion) transfer from the acid to the base. Two ions or molecules that *differ only by that one hydrogen ion* make up a **conjugate acid-base pair**. Three example pairs are shown below:

$$H_3O^+ \text{ and } H_2O \qquad H_2O \text{ and } OH^- \qquad NH_4^+ \text{ and } NH_3$$

The **conjugate acids** have one more proton (H^+) than the **conjugate bases**. For example, consider reaction (1) from Model 1:

$$\overbrace{CH_3COOH + H_2O}^{\text{conjugate acid-base pair}} \rightleftharpoons CH_3COO^- + H_3O^+$$

acetic acid + water $\qquad$ acetate + hydronium ion

conjugate acid-base pair

- The conjugate base is what results after the acid gives up a hydrogen ion; so we say that *acetate is the conjugate base of acetic acid*.
- The conjugate acid is what results after the base picks up a hydrogen ion; so we say that *hydronium ion is the conjugate acid of water*.

Critical Thinking Questions:

8. Consider reaction (2) in Model 1.

   a. What is the conjugate base of HCl?

   b. What is the conjugate acid of NH_3?

   c. Following the example in Model 2, draw lines onto reaction (2) in Model 1 to connect the conjugate acid-base pairs.

9. For the equations below, work with your team to label the acid and base on the reactant side and the conjugate acid and conjugate base on the product side. Draw a line to connect conjugate acid-base pairs together.

   a. $HCN(aq) + OH^-(aq) \rightleftharpoons H_2O(l) + CN^-(aq)$

   b. $F^-(aq) + H_3O^+(aq) \rightleftharpoons H_2O(aq) + HF(aq)$

Model 3: Strong and Weak Acids

When a generic acid (HA) is added to water, the following reaction results.

$$HA + H_2O \rightleftharpoons A^- + H_3O^+$$

The diagrams below represent the relative amounts of chemical species that are present in solutions of two hypothetical acids at equal concentration. Water molecules are so numerous that they are not shown in the diagrams. (The total volume of each solution is about 170 mL.)

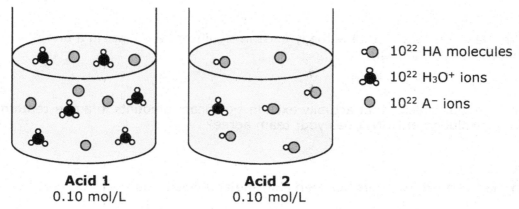

Acid 1
0.10 mol/L

Acid 2
0.10 mol/L

10^{22} HA molecules

10^{22} H_3O^+ ions

10^{22} A^- ions

Critical Thinking Questions:

10. Consider Model 3. List the chemical species (other than H_2O) that are present in an aqueous solution of Acid 1 (HA). Check that all team members agree.

11. List the chemical species (other than H_2O) that are present in an aqueous solution of Acid 2 (HA).

12. A **strong acid** is one that is essentially 100% dissociated in water, meaning it reacts completely with water, and the reaction goes 100% to products. Which acid (1 or 2) is a strong acid? Work with your team to explain why.

Model 4: Some common acids and bases

Type of electrolyte	Acids	Bases
Strong (all not listed here are weak)	HCl	LiOH
	HBr	NaOH
	HI	KOH
	H_2SO_4	$Ca(OH)_2$
	HNO_3	$Sr(OH)_2$
	$HClO_4$	$Ba(OH)_2$
Weak	$HC_2H_3O_2$	$Mg(OH)_2$
	HCN	

When weak acids and bases dissolve in water, they dissociate only <u>slightly</u> into ions.

CA30A

Critical Thinking Questions:

13. Write the three chemical species that actually exist in significant amounts in a one-tenth molar aqueous solution of HCl.

14. Work with your team to explain why hydrogen ion (H^+) is *not* one of the three species in CTQ 13. (Hint: Refer to the equation in Model 3.)

15. Write the three species that actually exist in significant amounts in a one-tenth molar aqueous solution of LiOH.

16. Write the four species that actually exist in significant amounts in a one-tenth molar aqueous solution of HCN. Does your team agree?

17. Summarize the differences between a strong and weak acid.

18. Cite an example of how you carried out your team role today.

Exercises:

1. What is the conjugate acid of HCO_3^-? _____ What is the conjugate base of HCO_3^-? _____
2. Is bicarbonate ion amphoteric? Explain.

3. Choose all the terms from the following list that apply to bicarbonate ion: **strong acid, strong base, weak acid, weak base**. Explain your choices.

4. Rewrite reaction (3) from Model 1 as the reverse reaction (that is, switch the reactants and products.) Draw lines to connect the conjugate acid-base pairs.

5. Write the species present in aqueous solutions of each of the following.
    a. NaOH
    b. HNO_2
    c. $Mg(OH)_2$
    d. HBr

6. Read the assigned pages in your textbook, and work the assigned problems.

pH
(In water, can hydronium exist without hydroxide?)

Model 1: Hydronium-hydroxide balance

In pure water, a small amount of self ionization occurs, with one water molecule acting as an acid (donating a proton) and another as a base (accepting a proton):

$$H_2O(l) + H_2O(l) \rightleftharpoons H_3O^+(aq) + OH^-(aq)$$

Further, the concentrations of hydronium ion and hydroxide ion in pure water are each 1.0×10^{-7} M, and the product at 25°C is always 1.0×10^{-14} M.

$$[H_3O^+][OH^-] = 1.0 \times 10^{-14}$$

Critical Thinking Questions:

1. Coffee has an $[H_3O^+]$ concentration of about 2.0×10^{-5} M. What is the $[OH^-]$ in coffee? Show your work.

2. Household ammonia has a hydroxide ion concentration of 4.3×10^{-3} M. What is the concentration of $[H_3O^+]$?

Model 2: pH

pH (the "power of hydrogen") is defined as the negative of the logarithm of the molar concentration of hydronium ions (without units):

$$pH = -\log[H_3O^+]$$

Therefore, in pure water:

$$pH = -\log(1.0 \times 10^{-7}) = 7.0$$

Critical Thinking Questions:

3. What is the pH of the coffee in CTQ 1?

4. What is the pH of the ammonia solution in CTQ 2?

5. Given the following $[H_3O^+]$ concentrations, calculate the pH values of these household solutions.

    a. lemon juice: $[H_3O^+] = 1.9 \times 10^{-2}$ M   _____

    b. apple cider: $[H_3O^+] = 5.0 \times 10^{-4}$ M   _____

    c. household cleaner: $[H_3O^+] = 6.3 \times 10^{-10}$ M   _____

6. As shown in Model 1, in a neutral solution $[H_3O^+] = [OH^-]$. What would be the pH of a neutral solution?

7. Most foods and beverages are acidic, while household cleaners such as ammonia are basic. Considering the pH of the solutions calculated in this activity, work with your team to predict the range of pH values that characterize:

   a. an acidic solution.

   b. a basic solution.

8. a. If the pH of a cola drink is 3.2, what is the hydronium ion concentration? Be sure that all team members can enter this into their calculator correctly, using the 10^x key (the 10^x key is often an inverse or 2nd log). Consult other teams or the instructor for help if necessary.

   b. What is the hydroxide ion concentration in the cola?

9. a. What is the hydroxide ion concentration in a 1.0×10^{-5} M aqueous solution of NaOH?

   b. What is the pH of this solution? (Careful!) Does your team agree?

10. What is the pH of a one-tenth molar solution of HCl?

    a. First, you must find the $[H_3O^+]$. Write the equation of HCl and water.

    b. Is HCl a strong or weak acid?_____

    c. Based on (b), what would be the concentration of $[H_3O^+]$? _____

    d. What is the pH? _____

11. List one or two problem solving strategies that your team employed today.

12. What is one goal to improve your team performance next time? How do you plan to accomplish it?

Exercises:

1. Calculate the pH of each of the following aqueous solutions.

    a. 1.0×10^{-4} M nitric acid

    b. 5.0×10^{-3} M hydrobromic acid

    c. a 1.0×10^{-6} M solution of the diprotic acid H_2SO_4 (diprotic means that each molecule of H_2SO_4 donates <u>two</u> hydrogen ions to water molecules)

    d. 0.0012 M $Ca(OH)_2$

2. Calculate the hydroxide ion concentrations of the solutions in Exercise 1.

3. Write the chemical formulas for the acids in Exercises 1a and 1b.

4. Write formulas for the conjugate bases of the acids in Exercises 1a and 1b.

5. Calculate the hydronium and hydroxide concentrations in each of the following aqueous solutions.

    a. Chicken broth, pH 5.80

    b. Blood plasma, pH 7.40

6. Read the assigned pages in your textbook, and work the assigned problems.

Acidity Constant (Ka)*
(How strong is it?)

Model 1: Keq and Ka for acids

When an acid, HA, is placed in water, hydronium ions are produced according to the reaction:

$$HA \text{ (aq)} + H_2O \text{ (l)} \rightleftharpoons H_3O^+ \text{ (aq)} + A^- \text{ (aq)}$$

The equilibrium constant, K_{eq}, for this type of reaction is:

$$K_{eq} = \frac{[H_3O^+][A^-]}{[HA][H_2O]}$$

Most solutions are sufficiently dilute that the concentration of water can be considered to be the same before and after reaction with the acid. The concentration of the water is incorporated into the value of K_{eq} and the equilibrium expression is given a special name and symbol: *the acid-dissociation constant, K_a*.

$$[H_2O] = 55.5 \text{ M}$$

$$K_{eq} = \frac{[H_3O^+][A^-]}{[HA][55.5]}$$

$$K_{eq} \times 55.5 = K_a = \frac{[H_3O^+][A^-]}{[HA]}$$

Table 1: Ka values for selected acids.

Acid Name	Molecular Formula	K_a value
hydroiodic acid	HI	3×10^9
hydrobromic acid	HBr	1×10^9
hydrochloric acid	HCl	1×10^6
perchloric acid	$HOClO_3$	1×10^8
sulfuric acid	H_2SO_4	1×10^3
hydronium ion	H_3O^+	55
nitric acid	HNO_3	28
acetic acid	CH_3COOH	1.75×10^{-5}
carbonic acid	H_2CO_3	4.5×10^{-7}
hydrofluoric acid	HF	7.2×10^{-4}
nitrous acid	HNO_2	5.1×10^{-4}
phosphoric acid	H_3PO_4	7.1×10^{-3}
water	H_2O	1.8×10^{-16}

The strength of an acid is determined by the relative H_3O^+ concentration produced at equilibrium for a given molarity of the acid. For example, if a 0.5 M solution of HA_1 has $[H_3O^+] = 1 \times 10^{-4}$ M and a 0.5 M solution of HA_2 has $[H_3O^+] = 1 \times 10^{-3}$ M, then HA_2 is a stronger acid than HA even though both are considered to be weak acids.

---

*Adapted from ChemActivity 43, Moog, R.S.; Farrell, J.J. *Chemistry: A Guided Inquiry*, 5th ed., Wiley, 2011, p. 256-263.

Critical Thinking Questions:

1. For HNO_3:

   a. Write a reaction with water.

   b. Write the K_a expression.

2. For an HNO_3 solution, $[H_3O^+] = [NO_3^-] = 0.967$ M. What is $[HNO_3]$ in this solution?

3. In a solution of nitrous acid: $[HNO_2] = 1.33$ M; $[H_3O^+] = 0.026$ M; $[NO_2^-] = 0.026$ M. Show that K_a is correct in Table 1.

4. Nitric acid and the acids above it in Table 1 are considered to be strong acids. Discuss with your team and be prepared to explain why to the class. (Recall the definition of strong acids from CA30A).

5. Acetic acids and the acids below it in Table 1 are considered to be weak acids. Work with your team and write a complete sentence to explain why.

6. Which of the weak acids in Table 1 (other than H_2O) will produce the highest $[H_3O^+]$ in a solution of a given molarity of acid? As a team, explain your reasoning. (You should not need to do extensive calculations).

7. Which of the weak acids in Table 1 (other than H_2O) will produce the lowest $[H_3O^+]$ in a solution of a given molarity of acid? As a team, explain your reasoning. (You should not need to do extensive calculations).

8. Rank the six acids in Table 1 with $K_a < 1$ in order from weakest to strongest. Be prepared to share your list with the class.

9. Given two acids and their respective K_a values, describe how you can determine which acid is stronger.

10. Did everyone in your team contribute to the activity today? If so, explain how. If not, identify what individuals need to do to ensure participation by all in the next session.

Exercises:

1. Write the chemical formulas for the conjugate bases for the acids listed in Table 1.

2. Write the balanced chemical equation for the reaction of HF with water. What is the expression for K_a? Choose one additional acid from Table 1 and provide the chemical reaction with water and the expression for K_a.

3. Indicate whether the following statement is true or false and explain your reasoning: A 0.25 M solution of acetic acid has a higher $[H_3O^+]$ than does a 0.25 M solution of nitrous acid.

4. Read the assigned pages in your textbook, and work the assigned problems.

Buffers
(How do acids and bases react together?)

Critical Thinking Questions:

1. Complete and balance the reaction of acetic acid donating a hydrogen ion to water to make acetate and hydronium.

$$HC_2H_3O_2 \quad + \quad H_2O \quad \rightleftharpoons$$

2. Draw lines connecting any conjugate acid-base pairs in the reaction in CTQ 1.

Model: Solutions ("systems") that may or may not be buffers.

ID	System obtained by mixing 100 mL of each of the two listed solutions	Initial pH	pH after addition of 100 mL of 0.200 M NaOH	pH after addition of 100 mL of 0.200 M HCl
1	0.100 M $HC_2H_3O_2$ and 0.100 M $NaC_2H_3O_2$	4.73	4.91	4.55
2	0.100 M $HC_2H_3O_2$ and 0.100 M NaCl	2.88	4.16	1.71
3	0.100 M $HC_2H_3O_2$ and 0.050 M $NaC_2H_3O_2$	4.43	4.67	4.12
4	0.100 M HCN and 0.100 M NaCN	9.21	9.39	9.03

Critical Thinking Questions:

3. Compare the solutions in the Model. List any systems (by ID number) that demonstrate large changes in pH (> 1 unit) when either acid or base is added.

4. List any systems that demonstrate small changes (<0.5 unit) in pH when either acid or base is added.

5. Work with your team to identify one or two characteristics that are shared by the solutions demonstrating small changes in pH but not by those demonstrating large changes in pH. Do not spend more than a minute or so on this task.

6. A **buffer** is a solution which is resistant to changes in pH when small amounts of acid or base are added. Based on this definition, which systems from the Model would make good buffers?

7. a. Use the molarity of acetic acid added to System (1) in the Model and the volume of the solution added (100 mL) to calculate the number of moles of acetic acid added to the system.

b. How many moles of acetate ion were added to this system? (Hint: sodium acetate and acetate ion have the same molar concentration).

8. Since System (1) contains both acetic acid and acetate ion, the equilibrium reaction you completed in CTQ 1 applies to this system. Suppose that some strong base (hydroxide ion) is added to this system. Write the two species from the reaction in CTQ 1 that could react with and *neutralize* the hydroxide. Work with your team to explain your choices.

9. Write a chemical equation for the reaction of hydroxide ion with acetic acid, producing acetate and water. Is hydroxide neutralized in this reaction? _____

10. As a team, explain why addition of strong base (hydroxide ion) to System (1) will not cause a great change in the pH of the solution, as long as the acetic acid is not used up.

11. What is the **one** species in CTQ 1 that could react with and neutralize any added hydronium ion?

12. Explain why addition of strong acid will not cause the pH of the solution in CTQ 1 to change much, as long as plenty of acetate is present.

13. Consider the following reaction:

$$HCl + H_2O \longrightarrow Cl^- + H_3O^+$$

a. Recalling that strong acids dissociate completely in water, draw a large 'X' through the species in the above reaction that will not be present in any significant amount.

b. Why is a forward arrow ($\longrightarrow$) used in this reaction instead of an equilibrium arrow ($\rightleftharpoons$)?

c. Work with your team to explain the following true statement: There is no species in the above reaction that can neutralize added hydronium ion.

14. Which would make a better buffer system: a strong acid and its conjugate base or a weak acid and its conjugate base? Explain.

15. Summarize the ingredients necessary for an effective buffer system.

16. How does the pH of a buffer system change when a small amount of acid is added?

17. How does the pH of a buffer system change when a small amount of base is added?

18. Which system in the Model is not an effective buffer?_____ Considering your answer to CTQ 16, what does this system lack that causes it to be ineffective at buffering the pH?

19. What is the most important unanswered question about buffers remaining with your team?

Exercises:

1. Systems (1) and (3) in the Model contain the same conjugate acid-base pair. When the same small amounts of acid or base are added to each system, which one exhibits a slightly larger pH change? _____ Explain what difference in these systems leads to this slight decrease in buffering ability.

2. State if each system would be useful as a buffer or not. Then explain the reason(s) for your choice.
   a. A solution containing 0.08 M NaCN and 0.10 M HCN

   b. A solution containing 0.05 M NaOH in H_2O

   c. A solution containing 0.25 M HCl and 0.20 M NaCl

   d. A solution containing 0.05 M NH_4Cl and 0.10 M NH_3

   e. A solution containing 0.20 M KF and 0.15 M HF

3. Considering the solution in Exercise 2b.
   a. Write the three chemical species that would be present in significant amounts.

   b. Is there any species present that can neutralize added hydroxide? Explain.

4. Read the assigned pages in your textbook, and work the assigned problems.

CA31A

Henderson-Hasselbalch Equation
(How do buffers prevent large pH changes?)

Model 1: pH of buffer systems

To calculate the pH of a buffer system, the *Henderson-Hasselbalch equation* is used.

$$pH = pKa + \log \frac{[A^-]}{[HA]}$$

Recall that HA represents any acid, and A^- represents the conjugate base.

Figure 1. Effect of acid and base addition to buffer and water.

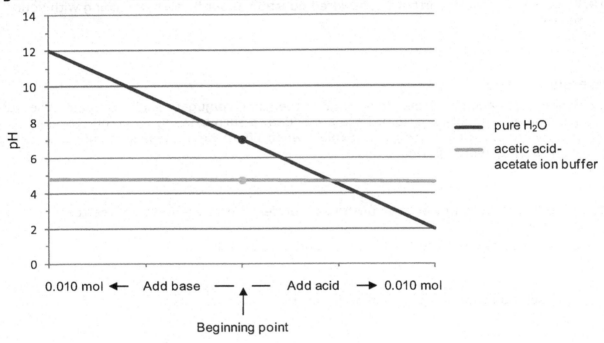

Critical Thinking Questions:

1. The K_a value for acetic acid is 1.85×10^{-5}. What is the pK_a of the buffer system in Figure 1?

2. Using the equation in Model 1, calculate the initial pH of the acetate buffer system if $[CH_3COOH] = [CH_3COO^-] = 0.10$ M.

Manager: Have your team work together to answer CTQ 3a-c. Ensure that every team member understands each part before moving on.

3. a. If 0.01 moles of strong acid are added to one liter of the acetic acid-acetate buffer from CTQ 2, what is the new concentration for total acid in the system [HA] (initial plus added amount)?

b. What is the new concentration of base, $[A^-]$? (Recall some of the base will be used to neutralize the added acid).

c. Using the Henderson-Hasselbalch equation, calculate the pH for this solution.

4. Using the Henderson-Hasselbalch equation, calculate the pH of the solution when 0.01 moles of base are added to one liter of the original solution. (Recall that some of the acid will be used to neutralize the added base).

5. **Manager:** Lead a team discussion comparing the changes in pH calculated for the buffer system in CTQs 3 and 4 to the same additions to water observed in Figure 1. Record your observations.

6. How does your answer to CTQ 5 help your understanding of the purpose of buffers?

7. Phosphoric acid (H_3PO_4) has a K_a value of 7.1×10^{-3}.

   Manager: Work with your team to first outline the necessary steps to solve the following problem.

   a. What is the pH of the buffer solution with initial concentrations of 0.45 M H_3PO_4 and 0.28M $H_2PO_4^-$?

   b. What is the new pH when 0.15 moles of base is added per liter of buffer solution?

CA31B

Model 2: Biological buffers

Buffers are vital for biological functions within the body. The carbonic acid – bicarbonate buffer maintains blood pH close to 7.4. Minor changes in pH can lead to coma and/or death. Carbonic acid (H_2CO_3) is unstable, causing the additional step in the following reaction. The lungs and kidneys help the system by removing excess CO_2 and acid, respectively.

Figure 2. Carbonic acid – bicarbonate buffer.

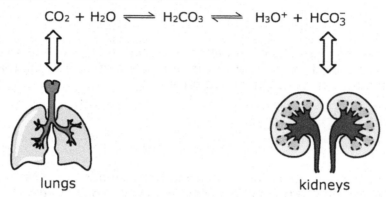

$$CO_2 + H_2O \rightleftharpoons H_2CO_3 \rightleftharpoons H_3O^+ + HCO_3^-$$

lungs   kidneys

(Note: For simplicity, H_2O was omitted as a reactant in the second step)

Critical Thinking Questions:

8. Write a K_a expression for the reaction in Figure 2. In order to do this, you can ignore H_2CO_3, considering the reactants to be CO_2 and H_2O and the products to be H_3O^+ and HCO_3^-.

9. Rearrange the equation in CTQ 8 to solve for $[H_3O^+]$.

Manager: *Have your team work together for CTQ 10. Ensure that every team member understands before moving on.*

10. a. What is the relationship between $[H_3O^+]$ and pH? That is, how does pH change when $[H_3O^+]$ is increased?

b. Consider your equation in CTQ 9. Work with your team to determine what would happen to $[H_3O^+]$ if the concentration of $[CO_2]$ increases? What would happen to the pH?

c. Explain your answer to (b) using Le Chatelier's Principle.

11. a. Write the Henderson-Hasselbalch equation for the second reaction in Figure 2.

b. Because the two reactions in Figure 2 share a compound, mathematically we can substitute $[CO_2]$ for $[H_2CO_3]$ in part (a). Rewrite the equation using this change.

12. a. The K_a for carbonic acid is 4.3×10^{-7}. What is the pH of a solution containing equal molar concentrations of HCO_3^- and CO_2?

b. Calculate the new pH if the concentration of CO_2 were twice that of HCO_3^-.

c. To maintain the usual pH of 7.4, would blood have equal molar concentrations of HCO_3^- and CO_2? If not, which would be higher? Work with your team to explain.

13. Some athletes consume sodium bicarbonate ($NaHCO_3$) before intense exercise to combat fatigue. Explain how this supplement (if effective) might help offset metabolic increases in CO_2 during exercise.

14. Work with your team to summarize how any increase or decrease in acid concentration can be minimized by the buffer system in the blood.

15. Give an example of how the Manager's role today helped ensure everyone understood the material.

16. What was one strength of your team today? Why was it beneficial for completion of the activity?

Exercises:

1. Formic acid (H_2COO) has a K_a of 1.8×10^{-4}. What is the pH of a buffer solution containing 0.32 M HCOOH and 0.14 M HCOO⁻?

2. Nitrous acid (HNO_2) has a K_a of 4.5×10^{-4}. What is the pH of a system containing 0.100 M nitrous acid and 0.100 M nitrite (NO_2^-) when 0.020 moles of NaOH is added per liter?

3. Excessive exercise can lead to production of lactic acid by the muscles and release into the bloodstream. Explain how the carbonic acid – bicarbonate buffer system would work to handle the increase in acid concentration to minimize changes in pH.

4. Read the assigned pages in your textbook, and work the assigned problems.

Titrations
(How can the concentration of a solution be determined?)

Model 1: The neutralization of acetic acid with sodium hydroxide

$$HC_2H_3O_2 \text{ (aq)} + NaOH \text{ (aq)} \rightleftharpoons NaC_2H_3O_2 \text{ (aq)} + H_2O \text{ (l)}$$

Critical Thinking Questions:

1. Is the reaction in Model 1 balanced?

2. What kind of double displacement reaction is shown in Model 1? _____

3. What are the chemical names of the two <u>reactants</u> in Model 1? The two <u>products</u>?

Model 2: pH of 40.0 mL of an acetic acid solution of unknown concentration after additions of 0.100 M sodium hydroxide (shown in both table and graph).

Volume of 0.100 M NaOH added, mL	pH
0	2.92
1	3.47
2	3.79
3	3.98
4	4.13
5	4.25
10	4.67
15	5.03
20	5.45
21	5.57
22	5.72
23	5.91
24	6.23
25	8.78
26	11.29
27	11.59
28	11.75
29	11.87

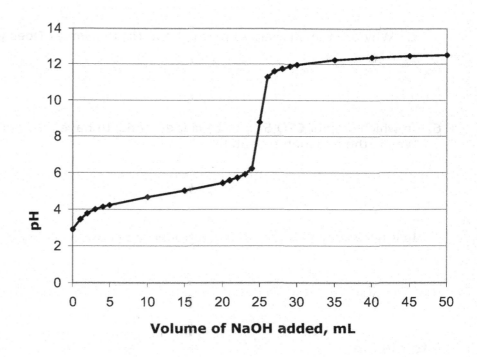

Critical Thinking Questions:

4. The equivalence point of a neutralization reaction can be identified by a steep rise in pH on both sides side of the data point.

    a. Circle the point on the graph in Model 2 that is at the equivalence point.

    b. What is the pH at the equivalence point of the titration in Model 2? _____

    c. How many mL of NaOH were added to the solution at the equivalence point? _____

    d. What was the concentration of NaOH added to the solution? _____

    e. Given your answers for (c) and (d), what quantity can you determine?

5. a. How many moles of NaOH were added at the equivalence point in Model 2?

    b. At the equivalence point, the base has exactly neutralized the acid and no extra base has been added. Considering your answer to part (a) and the equation in Model 1, how many moles of acetic acid ($HC_2H_3O_2$) reacted at the equivalence point?

    c. Why are the answers to part (a) and (b) the same? Does your team agree?

6. In which step of CTQ 5 (a or b) is it essential to have a balanced chemical equation? ____ Why is the equation essential?

7. Is it necessary that the pH is neutral at the equivalence point? Explain.

Information:

Often the exact concentration of an acid or base is unknown. To determine the concentration, a neutralization reaction known as a **titration** may be performed. An acid is titrated with a base when carefully measured volumes of a base of <u>known</u> molar concentration are added to a specific volume of an acid of <u>unknown</u> concentration.[*] The reaction is complete when the *equivalence point* is reached. **The equivalence point is the point at which all the acid present has been neutralized by added base and no additional base is added.**

_____

[*] A base can also be titrated by reacting with an acid of known molar concentration.

Critical Thinking Questions:

8. At the equivalence point in the reaction in Model 2, what two quantities are equivalent?

9. How many mL of $HC_2H_3O_2$ were titrated (that is, how many mL were there?)

10. What was the original molarity (that is, mol/L) of the acetic acid? Does your team agree?

11. Work with your team to summarize the steps necessary to solve titration problems.

12. Consider this titration problem: Suppose that 10.0 mL of 0.500 M NaOH were needed to reach the equivalence point when added to 15.0 mL of an acetic acid solution. Use the steps you outlined in CTQ 11 to calculate the molarity of the acetic acid solution. Work with your team to find the answer.

13. What is one improvement that your team can make for the next class period? How does your team plan to accomplish this?

Exercises:

1. Write the complete ionic equation and the net ionic equation for the reaction in Model 1.

2. A 25.0-mL sample of a KOH solution of unknown concentration required 66.4 mL of 0.150 M H_2SO_4 solution to reach the equivalence point in a titration.

   a. Write a balanced equation for the reaction.

   b. What is the molarity of the KOH solution?

3. An old bottle of aqueous HCl was found in the lab. Unfortunately, the concentration on the label was unreadable, so a titration was performed with 0.20 M NaOH.

   a. Write a balanced equation for the reaction.

   b. If 38.2 mL of 0.20 M NaOH was required to titrate a 15.0 mL sample of the acid, what is the HCl concentration?

4. How many milliliters of 1.50 M NaOH solution are required to titrate 40.0 mL of a 0.10 M HCl solution?

5. How many milliliters of 0.150 M NaOH are required to neutralize 50.0 mL of 0.200 M H_2SO_4?

6. Read the assigned pages in your textbook, and work the assigned problems.

Alkanes, Cycloalkanes and Alkyl Halides
(What makes a molecule "organic?")

Information:

Organic molecules are based on carbon backbone structures. Of these, hydrocarbons contain only carbon and hydrogen. The hydrocarbons containing no multiple bonds are called **alkanes**. The first ten "straight-chain" alkanes are shown in Table 1.

Table 1. Names and structures of the first ten alkanes (and alkyl groups)

Condensed Structure	Name of alkane	"Stick" Structure	Condensed Structure	Name of alkyl group substituent
CH_4	methane	none	$-CH_3$	methyl
CH_3CH_3	ethane	 ——	$-CH_2CH_3$	ethyl
$CH_3CH_2CH_3$	propane	 ∧	$-CH_2CH_2CH_3$	propyl
$CH_3CH_2CH_2CH_3$	butane	 ∧∨	$-CH_2CH_2CH_2CH_3$	butyl
$CH_3(CH_2)_3CH_3$	pentane	 ∧∨∧	$-CH_2(CH_2)_3CH_3$	
$CH_3(CH_2)_4CH_3$	hexane	 ∧∨∧∨	$-CH_2(CH_2)_4CH_3$	
$CH_3(CH_2)_5CH_3$	heptane		$-CH_2(CH_2)_5CH_3$	
$CH_3(CH_2)_6CH_3$	octane		$-CH_2(CH_2)_6CH_3$	
$CH_3(CH_2)_7CH_3$	nonane		$-CH_2(CH_2)_7CH_3$	
$CH_3(CH_2)_8CH_3$	decane		$-CH_2(CH_2)_8CH_3$	

The structures shown in the first column are called **condensed structures**. They can be written as a molecular formula by adding all the carbons and hydrogens (*e. g.*, ethane = C_2H_6) or expanded into a complete Lewis structure (*e. g.*, ethane, shown below):

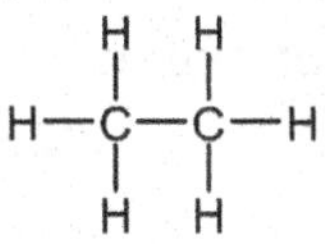

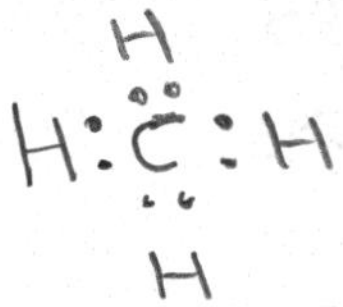

Critical Thinking Questions:

1. Draw a complete Lewis structure for propane.

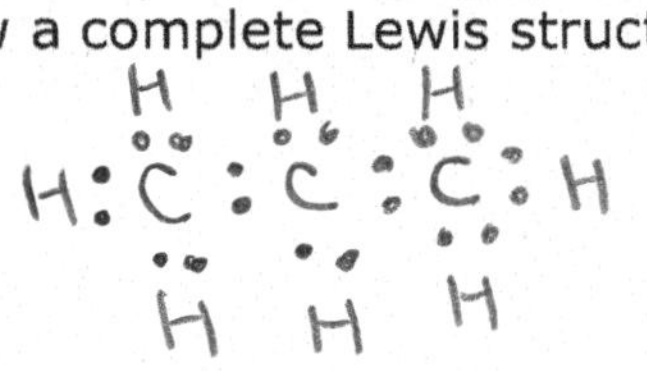

2. Complete Table 1 by filling in the missing "stick" **structures** and **names** of the alkyl group substituents. Ensure that all team members agree.

3. Circle **each** carbon atom in the stick structure of butane shown at the right. Which atoms are **not** shown in stick structures? **H** or **C** (circle one).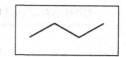

4. How does the structure of an alkyl group substituent differ from the structure of the alkane with the same number of carbons (see Table 1)?

5. Referring to Table 1, discuss with your team how the name of an alkyl group substituent can be derived from the name of the corresponding alkane. Write your team's answer in a complete, grammatically correct sentence.

Information:

Once organic molecules get large, it is convenient to draw them as "**stick structures**." Each "end" and each "bend" represents a carbon atom, and the hydrogens are not shown. Stick structures are commonly used for **cycloalkanes**—alkanes in which the carbons are connected to make a ring ("head" to "tail.") Some common cycloalkanes are shown in Table 2.

Table 2. Names and structures of some common cycloalkanes

Condensed Structure	Name of alkane	"Stick" Structure	Ball-and-Stick Model
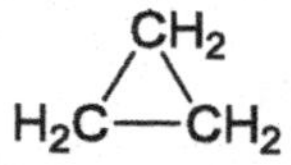 $H_2C\overset{CH_2}{\diagup}\diagdown CH_2$	cyclopropane	 △	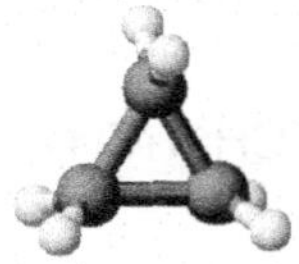
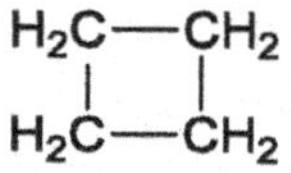 $\begin{array}{l}H_2C-CH_2\\H_2C-CH_2\end{array}$	cyclobutane	□	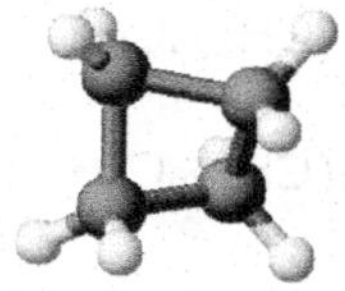
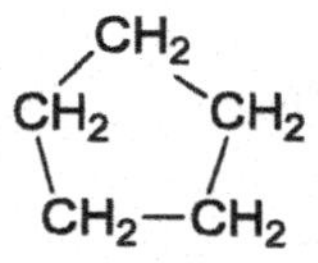 CH_2 CH_2 CH_2 CH_2-CH_2	cyclopentane	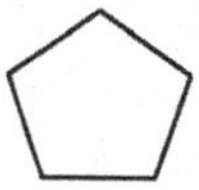 ⬠	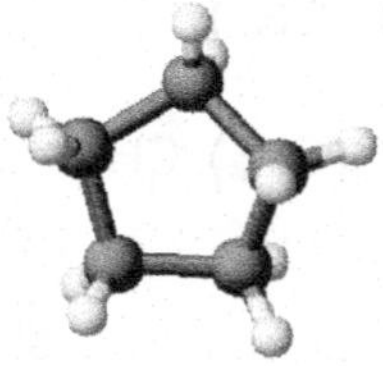
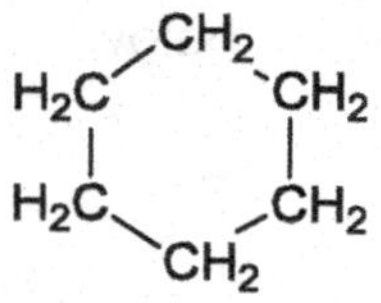 $H_2C\overset{CH_2}{\diagup}CH_2$ $H_2C\underset{CH_2}{\diagdown}CH_2$	cyclohexane	 ⬡	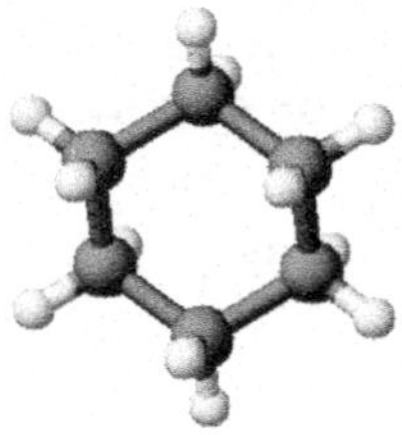

Critical Thinking Question:

6. By extension from Table 2, draw condensed and "stick" structures for cycloheptane.

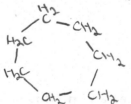

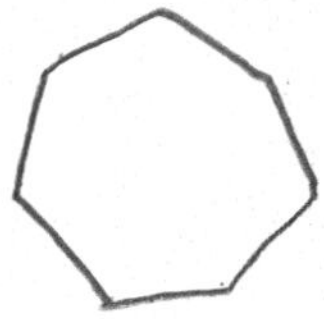

Information: Branched alkanes

Alkanes are **branched** when the carbons are not connected in a continuous manner. For example, if you cannot trace all the carbons with your pencil (or finger) from one carbon to the next without lifting your pencil or retracing a portion, there is a branch.

Table 3. Some names and structures of some substituted butanes

Structure	Name of alkane	Structure	Name of alkane
	2-methylbutane		2,2-dimethylbutane
	2-methylbutane		2,2-dimethylbutane
CH₃—CH₂—CH—CH₃ ⎮ CH₃	2-methylbutane		2,3-dimethylbutane
	2,2,3-trimethylbutane		2,2,3,3-tetramethylbutane

Critical Thinking Questions:

7. A circle has been drawn around the longest continuous carbon chain in the first molecule of Table 3. Draw a circle around the longest chain for the other 7 molecules in the table. (If there are two or more chains of the same length, draw the circle around any one you like.) How many carbons are in each of your circles? _____

8. "Butane" is the **base name** for each alkane in Table 3. Discuss with your team and complete the definition for the term **base name**:

   The **base name** of a branched alkane is _____

   _____

9. Now, draw a **box** around each branch that extends outside the circles you drew in Table 3. For example, the first molecule listed should look like the picture at the right. How many carbons are in each of your boxes? _____ How many hydrogens are connected to the carbon in each box? _____

10. What is the name of the **alkyl substituent** in each **box** you drew in CTQ 9? Refer to Table 1 if necessary.

11. What is indicated by the prefixes "di," "tri," and "tetra" on some of the names in Table 3?

12. The first three structures in the left column are all named "2-methylbutane." What is indicated by the number "2" in these names?

13. Explain why three numbers are needed to name the molecule "2,2,3-trimethylbutane."

14. What punctuation mark is used in names between two numbers? _____

15. What punctuation mark is used in names to join a number and a letter? _____

Information: Some rules for naming branched alkanes with multiple substituents

1. Find the longest continuous carbon chain; if two chains have the same length but different branching, choose the one with more branches.
2. Mentally number each carbon of the base chain starting at the end closest to the branch (or if both ends are equally close, choose the one closer to more branches.)
3. Name each different substituent in alphabetical order. If there are two equivalent substituents, add the prefixes di-, tri-, tetra, *etc.* in front of them.
4. Finally, give each substituent a number that corresponds to the number of the carbon of the base chain to which it is connected.

Table 4. Some names and structures of some branched alkanes

Structure	Name of alkane	Structure	Name of alkane
	3-ethyl-2-methylpentane		3-ethyl-2,4-dimethylpentane

Critical Thinking Questions:

16. Why is "3-methylbutane" an incorrect name for the structure below? (Hint: Which rule is violated?)

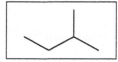

17. The 5-carbon chain circled in the structure below, left, is **not** a correct one. In the second drawing, circle a correct 5-carbon base chain. (Hint: see Rule 1 above.)

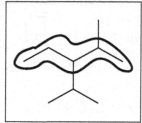 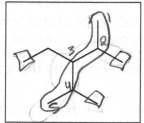

18. Following each rule in order, give the correct name for the molecule in CTQ 17.

    3-ethyl-2,4-dimethyl pentane

Information: Branched substituents and halogens

Sometimes the substituents themselves are branched. Some common substituent names are shown in Table 5. If the substituent is a halogen, then the molecule is called a **haloalkane** or **alkyl halide**.

methyl — CH_3 ethyl — CH_2CH_3 Propyl — $CH_2CH_2CH_3$

Table 5. Common substituent names in alkanes and haloalkanes. "R" indicates where the substituent is attached to the "Rest" of the molecule (the main chain).*

Substituent	Name	Stick structure	Example	Name
—$CHCH_3$ $\mid$ CH_3	isopropyl			4-isopropylheptane
—CH_2CH—CH_3 $\mid$ CH_3	isobutyl	R	Br	1-bromo-2-methylpropane (also called isobutyl bromide)
—$CHCH_2CH_3$ $\mid$ CH_3	sec-butyl	R		sec-butylcyclopentane
CH_3 $\mid$ —C—CH_3 $\mid$ CH_3	tert-butyl	R		tert-butylcyclohexane
—F	fluoro	R-F	CF_4	tetrafluoromethane
—Cl	chloro	R-Cl	$ClCH_2CH_2Cl$	1,2-dichloroethane
—Br	bromo	R-Br	Br	2-bromobutane (also called sec-butyl bromide)
—I	iodo	R-I	CHI_3	triiodomethane

* The prefix "iso" means "isomer," which in this case indicates different branching. So, a propyl group is not branched, but an isopropyl group is. The prefixes sec- and tert- stand for secondary and tertiary, respectively, and can be abbreviated with the single letters s- and t-. We will learn more about how these names are derived later in the course.

Critical Thinking Questions:

19. Draw the molecule implied by the name 2-methyl-3-propylpentane. Looking at the structure you drew, explain why the name is incorrect. (Hint: Apply each naming rule in order to the structure that you drew.) Write the correct name.

CH_3—CH—CH—CH_2—CH_3
        $\mid$ $\mid$
       CH_3 CH_2—CH_2—CH_3

3-ethyl-2-methyl hexane

20. List two to three "helpful hints" that your team found important for naming and drawing compounds.

Drawing the structure

Using table 5

Naming branches in Alphabetical order

Counting each point in a line angle

21. Write one strength of your teamwork today and why it was important.

Exercises:

1. Molecules that have the same molecular formula but are arranged (*e. g.*, branched) differently are called **isomers**. Draw (any style) and name all 5 of the differently branched isomers of C_6H_{14}.

2. Complete the following table:

Name	Lewis structure	condensed formula	stick structure
2-chloropropane	H $\overset{Cl}{\underset{\mid}{C}}$ H $\quad$ H—C—C—C—H $\quad$ H H H	$\overset{Cl}{\underset{\mid}{CH_3CH}} CH_3$	$\overset{Cl}{\diagup}$
		$\underset{CH_3CH_2CH-CH-CH-CH_3}{\overset{CH_3 \qquad CH_2CH_2CH_3}{\mid \qquad\qquad \mid}}$ $\underset{CH_2CH_3}{\mid}$	
isopropylcyclohexane			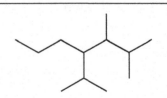

3. Read the assigned pages in your textbook and work the assigned problems.

Conformers
(How and why do molecules "twist?")

Model 1: Representations of ethane in its most favorable conformation

Ball-and-stick model | Spacefilling model | Lewis dash-and-wedge structure | Newman projection

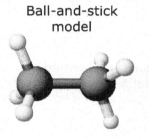 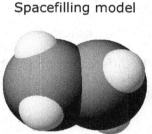 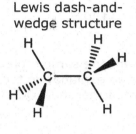 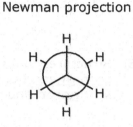

Critical Thinking Questions:

1. Each representation in Figure 1 shows ethane (CH_3CH_3) in its lowest potential energy (most favorable) conformation. If you have a model set available, make a model of ethane and rotate the single bonds until it is in this conformation.

   a. Construct an explanation for why this conformation is the most favorable.

   b. Consider the Newman Projection in Figure 1. What atom is at the center of the picture?

   c. The atom you named in CTQ 1b is represented as a large circle or disc. What single atom is hidden from view behind the disc?

2. Consider the Newman projection shown below of ethane in (nearly) its least favorable conformation. If you have a model set available, rotate the single bonds until it is in this conformation. Draw a dash-and-wedge structure for this conformation.

3. In your own words, explain what the term *conformation* means, as applied to ethane.

4. Look at your model "end-on" and compare with the Newman projections below. The angle you observe between the hydrogens at the top of each structure in **boldface** is called the torsional angle. Circle the correct torsional angle for each conformer.

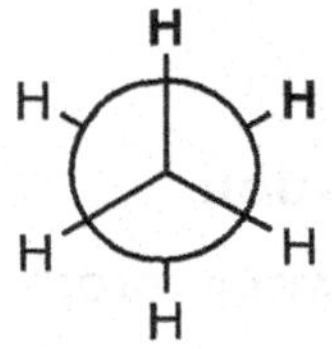

	Torsional angle (staggered)
	0°
	60°
	120°
staggered	180°

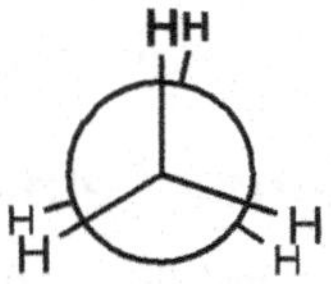

	Torsional angle (eclipsed)
	0°
	60°
	120°
eclipsed	180°

5. Structures that represent the same molecule in different conformations are termed <u>conformers</u>. Construct an explanation for why the conformer in Model 1 is called **staggered** and the conformer in CTQ 2 is called **eclipsed**.

6. What repulsive forces will cause ethane in the eclipsed conformation to quickly adopt the staggered conformation?

Model 2: Some representations of alkanes with 2, 3 and 4 carbons

Skeletal or "Stick" structure	Ball-and-stick structure	Lewis structure	Dash-and-wedge structure

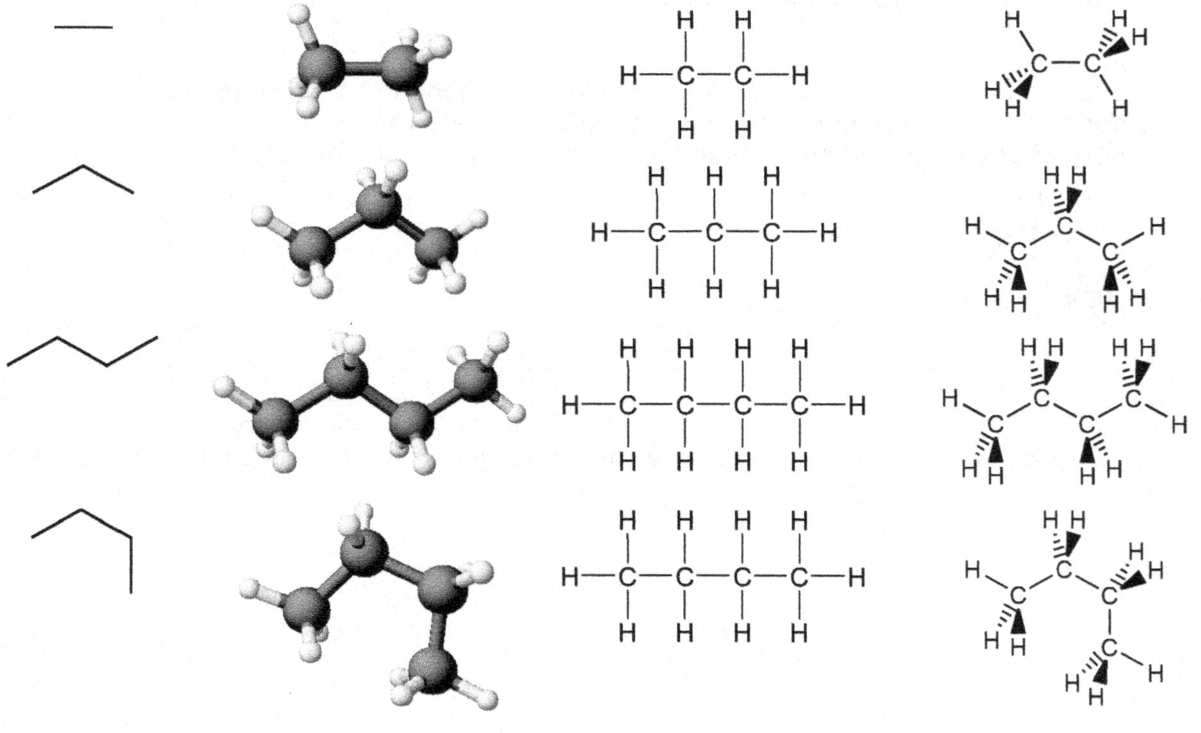

Critical Thinking Questions:

7. Underneath each stick structure in Model 2, write the name of the alkane.

8. The ball-and-stick structure below matches one of the two butane conformations shown in Model 2, looking "end-on" down the bond between carbons 2 and 3. Circle the stick structure in Model 2 that matches the structure below.

9. Complete the Newman projection at the right in CTQ 8 by adding H or CH₃ groups so that it represents the conformation shown in the ball-and-stick structure at the left.

10. Consider the <u>other</u> stick structure for butane in Model 2 (the one you did <u>not</u> circle in CTQ 8), sighting down the C2–C3 bond.

    a. Is this structure **staggered** or **eclipsed** (circle one)? You may use your model to help you.

    b. Draw a Newman projection for this conformation of butane.

    c. At room temperature, the single bonds in butane are continually rotating through the staggered and eclipsed conformations. However, the molecule spends more time closer to one of the two extremes. Which conformation is it likely to spend more time in— **staggered** or **eclipsed** (circle one)?

Model 3: Some conformations of pentane

    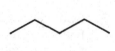

Critical Thinking Questions:

11. Circle the two structures of pentane in Model 3 that are in their most favorable conformation. Explain why these are the most favorable.

12. Draw a line to connect the two structures that you circled in Model 3. These two structures are in the same conformation, but the molecule as a whole is rotated.

13. Find two other structures in Model 3 that are identical (*i. e.,* in the same conformation), and draw a second line to connect them.

14. What is the total number of distinct conformers of pentane shown in Model 3? _____

15. List one or two remaining questions from today's activity.

16. What are the main concept(s) for this activity?

Exercises:

1. Draw a wedge and dash-bond representation of pentane in its most favorable conformation.

2. Consider the molecule 2-methylbutane.

    a. Using the templates at the right, complete two staggered Newman projections for 2-methylbutane: one sighting down the C1–C2 bond and the second sighting down the C2–C3 bond.

    b. Complete the eclipsed Newman projection for 2-methylbutane sighting down the C2–C3 bond.

3. Sugars and other complex molecules are often depicted using a representation called a Fisher projection. In a Fisher projection all <u>horizontal</u> bonds are assumed to come out of the page toward you (wedge bonds) and all <u>vertical</u> bonds are assumed to go back into the page, away from you (dash bonds). Draw a wedge and dash representation of the Fisher projection of glyceraldehyde shown below.

4. Read the assigned sections in your text and work the assigned problems.

Constitutional and Geometric Isomers
(Are they identical, or are they isomers?)

Model 1: Representations of some organic molecules

Skeletal or "Stick" structure	Lewis structure	Ball-and-stick structure

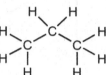

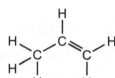

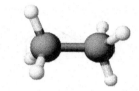

Critical Thinking Questions:

1. Consider the Lewis structures in Model 1. How many covalent bonds does each carbon have? _710, 9_

2. In skeletal representations, the hydrogens are not shown. Explain how it is still possible to tell how many hydrogen atoms there are on each carbon.

   each Point on the skeletal represents one C. Therefore, we attach the H to each C to complete the 4 bonds that each C needs

3. Draw a Lewis structure representation of the molecule for which a skeletal representation is shown below.

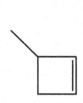

   H
   |
   H—C—H   H
           |
           C—C—H
        | ||
   H—C—C—H
        |
        H

CA35

Model 2: Constitutional Isomers

Column 1		Column 2	
structure	molecular formula	structure	molecular formula

Column 1:
- structure — C_5H_{12}
- structure — C_5H_{12}
- structure — C_5H_{12}

Column 2:
- structure — C_5H_{10}
- structure — C_5H_{10}
- structure — C_5H_{10}

Critical Thinking Questions:

4. Complete Model 2 by writing in the missing molecular formulas in both columns.

5. What do the molecules in a given column (1 or 2 in Model 2) have in common with the other molecules in that column?

   In each colum they have the same molecular formula

6. What do the molecules in a given column **not** have in common with the other molecules in that column?

   They differ in their structure

7. All the structures in a given column are **constitutional isomers** of one another, but the structures in Column 1 are not constitutional isomers of structures in Column 2. Based on this information, write a definition for the term **constitutional isomers**.

8. If the molecule shown below were placed into Model 2, would it belong in **Column 1** or **Column 2** (circle one)? Explain your choice.

Model 3: Representations of methylcyclobutane

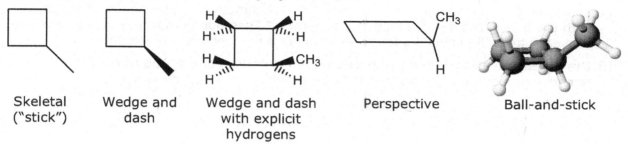

| Skeletal ("stick") | Wedge and dash | Wedge and dash with explicit hydrogens | Perspective | Ball-and-stick |

Model 4: 1,2-dimethylcyclobutane, shown with ring carbons numbered 1–4

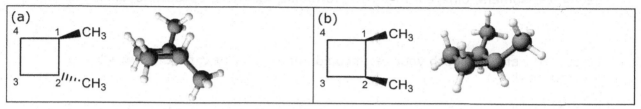

Critical Thinking Questions:

9. Do the molecules in boxes (a) and (b) of Model 4 meet your definition of constitutional isomers from CTQ 7? Explain. *Same molecular structure and have n different order of attachement*

10. Carbon 1 in box (a) is bonded to two carbons in the ring, a methyl group (–CH₃) and a hydrogen. *Other than bonds to carbons within the ring*, what two groups are bonded to the following carbons?

    a. carbon 1 in box (b)?

    b. carbon 2 in box (a)?

    c. carbon 2 in box (b)?

11. If you have access to a model kit, make models of the two molecules in Model 4 (C = black; H = white; use the short bonds for single bonds). Is it possible to twist single bonds in the models such that the molecule in box (a) is the same as the one box (b)?

Information:

Since each carbon in the molecule in box (a) in Model 4 is bonded to the same four groups as the corresponding carbon in the molecule in box (b), the molecules are said to have the same **connectivity**.

You confirmed in CTQ 11 that the two structures of 1,2-dimethylcyclobutane shown above are not simply **conformers** of each other. We examined conformers in ChemActivity 34.

Imagine that the four carbons of the cyclobutane ring define a plane. In one structure, the two methyl groups are on the *same side* of this plane, and in the other they are on *opposite sides* of the plane. The single bonds in the ring cannot rotate without breaking the ring. Two groups on the *same side* of the plane are considered to be *cis* to one another. Groups on opposite sides are called *trans*.

Geometric isomers (*cis-trans* isomers) are molecules that have the same <u>connectivity</u> and differ only in the <u>geometric arrangement</u> of groups.

Critical Thinking Questions:

12. Label one box in Model 4 with the name "*cis*-1,2-dimethylcyclobutane" and the other with the name "*trans*-1,2-dimethylcyclobutane." Then add perspective representations into each box. (See Model 3 for an example of a perspective representation.)

13. Draw wedge-and-dash and perspective representations of *cis*- and *trans*-1,3-dimethylcyclobutane. (Note: that is "1,3-dimethyl," not "1,2-dimethyl.")

14. Arrange the following descriptive terms in order from *most to least similar:* conformers, geometric isomers, different molecular formulas, identical, constitutional isomers.

Identical, Conformers, geometric isomers, Constitutional isomers, different molecular formulas

15. What is one area in which your team could improve for next time? How will you accomplish this?

Exercises: due wednesday

1. Indicate if the following pairs of structures are *identical, conformers, geometric isomers, constitutional isomers,* or *not isomers*.

a.	1,2-dimethylcyclobutane and 1,3-dimethylcyclobutane Constitutional	
b.	___/⟍ and ⟍=/⟍ C₅H₁₀ Constitutional C₅H₁₀ CH₃CH = CHCH₂CH₃ CH₃C=CHCH₃ CH₃	
c.	___/⟍ and ⟍=/⟍ CH₃CH=CHCH₂CH₃ Conformer C-C=C-C-C	
d.	and geometric C	
e.	and identical	
f.	and constitutional	
g.	and constitutional	
h.	and geometric	
i.	and not isomers	
j.	and not isomers	
k.	and constitutional	

2. Draw a structure for a molecule not shown in this activity that would belong in Column 2 of Model 2.

3. Read the assigned pages in your textbook and work the assigned problems.

CA35

Functional Groups

(How are organic molecules classified?)

Information:

Organic molecules can be thought of in terms of **functional groups**—certain arrangements of bonded atoms that have predictable properties and reactivities.

The tables below list some functional groups that are important in organic chemistry and biochemistry. Table 1 contains those which we will name by the systematic rules agreed upon by the International Union of Pure and Applied Chemistry (IUPAC). These rules assume that the most important functional group is assigned a **base name**, and other functional groups are named as **substituents**. Most functional groups have separate rules depending on whether it is the base name (named last) or the substituent (named first); these differences are indicated in the columns "base suffix" and "name when substituent" in the table.

Table 2 contains some functional groups which are easier to name by common (or "trivial") naming rules.

Sometimes more than one name is given for an example because IUPAC has adopted some common names as acceptable alternates. For example, the molecule *formaldehyde* is systematically named *methanal*. The second (systematic) name is more descriptive, but almost nobody uses it. You should be familiar with both names.

Table 1: Organic Functional Groups (Systematic IUPAC Naming Rules)

functional group	generic structure	base suffix	name when substituent	examples
alkane	R	-ane	-yl	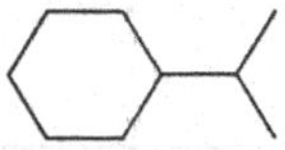 $CH_3 CHCH_3$ / CH_3 2-methylpropane isopropylcyclohexane
alkene	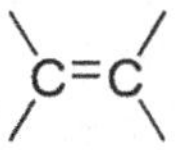 $\C=C\$	-ene	-enyl	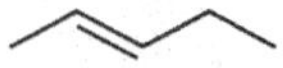 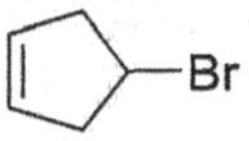 *trans*-2-pentene 4-bromocyclopentene (Br)
alkyne	$—C≡C—$	-yne	-ynyl	$CH_3 C≡C-CHCH_3$ / CH_3 4-methyl-2-pentyne
aromatic	 (benzene ring)	benzene	phenyl-	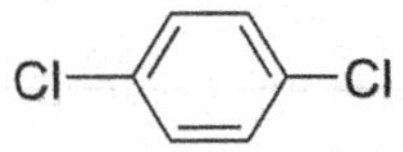 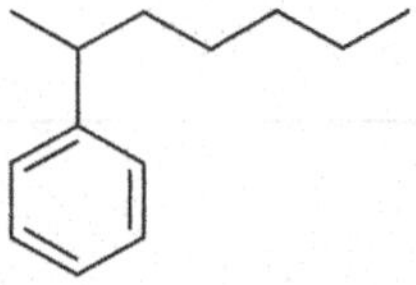 Cl——Cl 1,4-dichlorobenzene 2-phenylheptane
haloalkane	R-X	—	fluoro-, chloro- etc.	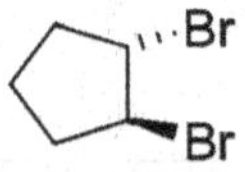 CF_2Cl_2 dichlorodifluoromethane *trans*-1,2-dibromocyclopentane (Br, Br)

Table 1 (continued: Organic Functional Groups (Systematic IUPAC Naming Rules)

functional group	generic structure	base suffix	name when substituent	examples
alcohol	R-OH	-ol	hydroxy-	CH_3CHCH_3 with OH, 2-propanol — $CH_3CHC{-}OH$ with O and OH, 2-hydroxypropanoic acid
phenol	benzene ring-OH	phenol	—	2,5-dichlorophenol (ring with OH, two Cl)
thiol	R-SH	-thiol	mercapto-	$HS{-}CH_2CH_2CH_3$ 1-propanethiol — $HSCH_2CH_2OH$ 2-mercaptoethanol
aldehyde	$R{-}\overset{O}{\underset{}{C}}{-}H$	-al	—	$H{-}\overset{O}{C}{-}H$ methanal (formaldehyde); benzaldehyde; CH_3CHCH with O and Cl, 2-chloro-propanal
ketone	$R{-}\overset{O}{\underset{}{C}}{-}R$	-one	—	$CH_3{-}\overset{O}{C}{-}CH_3$ 2-propanone (acetone); $CH_3{-}\overset{O}{C}{-}CHCH_3$ with CH_3, 3-methyl-2-butanone (methyl isopropyl ketone)
carboxylic acid	$R{-}\overset{O}{\underset{}{C}}{-}OH$	-oic acid	—	$CH_3{-}\overset{O}{C}{-}OH$ ethanoic acid (acetic acid); benzoic acid
acid salt	$R{-}\overset{O}{\underset{}{C}}{-}O^{\ominus}$	-oate	—	$CH_3{-}\overset{O}{C}{-}O^{\ominus}Na^{\oplus}$ sodium ethanoate (sodium acetate); benzoate
amide	$R{-}\overset{O}{\underset{}{C}}{-}NR_2$	-amide	—	$CH_3{-}\overset{O}{C}{-}NH_2$ ethanamide (acetamide); N,N-dimethylbenzamide

Table 2: Organic Functional Groups (Common Naming Rules)

functional group	generic structure	naming rules	examples	
ester	$R-\overset{\overset{\displaystyle O}{\|}}{C}-OR'$	name R' as alkyl group; then like acid salt	$CH_3CH_2\overset{\overset{\displaystyle O}{\|}}{C}-OCH_3$ methyl propanoate	isopropyl benzoate
acid anhydride	$R-\overset{\overset{\displaystyle O}{\|}}{C}-O-\overset{\overset{\displaystyle O}{\|}}{C}-R$	name each $R-\overset{\overset{\displaystyle O}{\|}}{C}-$ as acid alphabetically; then "anhydride"	$CH_3CH_2\overset{\overset{\displaystyle O}{\|}}{C}-O-\overset{\overset{\displaystyle O}{\|}}{C}-CH_2CH_2CH_3$ butanoic propanoic anhydride	$CH_3-\overset{\overset{\displaystyle O}{\|}}{C}-O-\overset{\overset{\displaystyle O}{\|}}{C}-CH_3$ ethanoic anhydride (acetic anhydride)
ether	R-O-R	name each R as alkyl group alphabetically; then "ether"	$CH_3CH_2OCH_2CH_3$ diethyl ether	methyl phenyl ether benzyl methyl ether
sulfide	R-S-R	name each R as alkyl group alphabetically; then "sulfide"	$CH_3CH_2SCH_2CH_3$ diethyl sulfide	
disulfide	R-S-S-R	name each R as alkyl group alphabetically; then "sulfide"	$CH_3CH_2SSCH_2CH_3$ diethyl disulfide	
amine	$R-\overset{\displaystyle ..}{\underset{\underset{\displaystyle R}{\|}}{N}}-R$	name each R as alkyl group alphabetically; then "amine"	CH_3NH_2 methyl amine	$(CH_3)_2NCH_2CH_3$ ethyl dimethyl amine
amino group	$-NH_2$	amino- prefix	$CH_3\underset{\underset{\displaystyle NH_2}{\|}}{C}H\overset{\overset{\displaystyle O}{\|}}{C}-OH$ 2-aminopropanoic acid	$HOCH_2\underset{\underset{\displaystyle NH_2}{\|}}{C}H-CH_3$ 2-amino-1-propanol

Notes:

- Several functional groups contain a carbonyl group ($-\overset{\overset{\displaystyle O}{\|}}{C}-$); the carbonyl is not considered a functional group by itself.

- When two or more functional groups are present, choose the more important one to be the **base name**. In general, the more carbon-oxygen bonds contained by a functional group, the higher its importance.

Exercises:

1. Identify the functional group that corresponds to the following descriptions.

   a. A carbonyl group connected to at least one hydrogen.   Aldehide

   b. An oxygen bonded to two saturated carbons.   ether

   c. A benzene ring bonded to a hydroxy group.   Phenol

   d. A carbon bonded to two other carbons and double-bonded to an oxygen.   Ketone

2. List all the functional groups (other than alkane) in each of the following structures.

   a. $CH_3CH_2-\overset{O}{\overset{\|}{C}}-CH_2CH_3$ Ketone

   b. 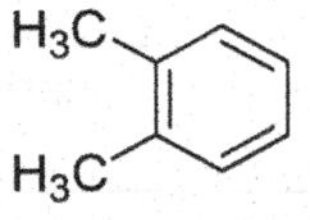 (H₃C / H₃C benzene structure)   Benzene ring   Aromatic

   c. $Br-CH_2\overset{O}{\overset{\|}{C}}-OH$ Carboxylic Acid

   d. $H_2N-\overset{O}{\overset{\|}{C}}-CH_2CHCH_3$ (CH₃)   Amide

   Prefix = #C

   infix - type of C-C bond

   e. 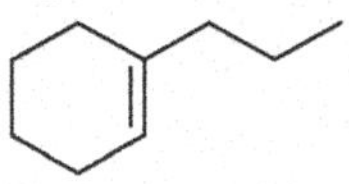 (cyclohexene with propyl)   Alkene

   f. $H_3C-\overset{CH_3}{\underset{CH_3}{\overset{|}{\underset{|}{C}}}}-OH$ Alcohol

   Suffix - Class functional

   g. $CH_3\overset{OH}{\overset{|}{C}}HCH_2-\overset{O}{\overset{\|}{C}}-O^{\ominus}$ Alcohol and acid salt

3. Give a correct (common or systematic) name for each molecule in Exercise 2.

   3,
   a. Pentanone

   b.

   c. ethanoic acid   2 bromoethanoic acid

   d. 3-methbutanamide

   e. propyl Cyclohexene

   f. 2 methyl Propanol

4. Draw the structure that corresponds to each of the following names.

   a. diisopropyl disulfide

b. butyl methyl ether

c. 2-methyl-2-aminopropanal

d. benzoic anhydride

e. 2,4,6-trimethylphenol

f. pentyl amine

g. 2-mercaptoacetic acid

5. List all the functional groups (other than alkane) in each molecule named in Exercise 4.

a.

b.

c.

d.

e.

f.

g.

6. Read the assigned sections in your text and work the assigned problems.

Overview of Organic Reactions
(What are the main types of organic reactions in biological systems?)

Information: Organic reactions common in biochemistry

There are seven common reaction classes of organic molecules that we will consider. The first class (acid-base reactions) is one we have seen before. Then there are three pairs of reactions that are opposites of each other—addition and elimination; reduction and oxidation; condensation and hydrolysis. Sometimes we do not write all the inorganic molecules, so the reactions as written are not balanced.

Table 1: Seven common types of reactions with biomolecules

1. **acid-base**: transfer of a proton (H^+)

   - example:

2. **addition**: addition of a small molecule (usually H_2O) across a <u>double bond</u>

   - example (hydration):
     $$CH_3-CH=CH_2 \ + \ H_2O \xrightarrow{\text{H}^+ \text{ catalyst}} CH_3-CH-CH_2$$
     with H and OH below

3. **elimination**: removal of a small molecule (often H_2O) to form a double bond

   - example (dehydration):
     $$CH_3-CH-CH_2 \xrightarrow[\text{and heat}]{\text{concentrated acid}} CH_3-CH=CH_2 \ + \ H_2O$$
     with H and OH below

4. **reduction**: addition of 2 H atoms to <u>or</u> removal of an O atom from a molecule. Can be symbolized "[r]"

   - example:
     $$CH_3-CH=CH_2 \ + \ H_2 \xrightarrow{\text{Pt catalyst}} CH_3-CH_2-CH_3$$

5. **oxidation**: addition of an O atom to <u>or</u> removal of 2 H atoms from a molecule. Can be symbolized "[o]"

   - example with primary alcohol: **first** forms an aldehyde, **then** forms an acid

     ethanol $\xrightarrow[(-H_2)]{[o]}$ acetaldehyde (ethanal) $\xrightarrow[(+O)]{[o]}$ acetic acid (ethanoic acid)

   - example with secondary alcohol: forms a ketone

     2-propanol $\xrightarrow[(-H_2)]{[o]}$ 2-propanone (acetone)

- example with tertiary alcohol: no reaction

$$H_3C-\underset{\underset{CH_3}{|}}{\overset{\overset{OH}{|}}{C}}-CH_3 \xrightarrow{[o]} NR \text{ (no H's on alcohol carbon to be removed)}$$

tert-butyl alcohol

- example with thiol: forms a disulfide

$$2 \, CH_3\text{-}SH \xrightarrow{[o]} CH_3\text{-}S\text{-}S\text{-}CH_3 \, (+ \, H_2O)$$

methanethiol   dimethyl disulfide

6. **condensation**: coupling of two molecules with the loss of a small molecule (usually H_2O)

- example (esterification):

$$CH_3-\overset{\overset{O}{||}}{C}-\boxed{OH} + \boxed{H}O\text{-}CH_3 \underset{}{\overset{H^+ \text{ catalyst}}{\rightleftharpoons}} CH_3-\overset{\overset{O}{||}}{C}-OCH_3 + H_2O$$

acetic acid   methanol   methyl acetate
(an acid)   (an alcohol)   (an ester)

7. **hydrolysis**: splitting a molecule in two with the addition of water

- example (ester hydrolysis):

$$CH_3-\overset{\overset{O}{||}}{C}-OCH_3 + H_2O \underset{}{\overset{\text{acid or base catalyst}}{\rightleftharpoons}} CH_3-\overset{\overset{O}{||}}{C}-\boxed{OH} + \boxed{H}O\text{-}CH_3$$

methyl acetate   new water added

Critical Thinking Questions:

1. For the acid-base reaction in Table 1, identify the acid and base on the reactant side and the conjugate acid and conjugate base on the product side. Draw a line to connect conjugate acid-base pairs together. Check that all team members agree.

2. Referring to Table 1, discuss with your team and choose which of the seven common reaction types is illustrated by each of the following.

a.
$$CH_3CH_2-\overset{\overset{O}{||}}{C}-CH_2CH_3 + H_2 \xrightarrow{Pt} CH_3CH_2-\underset{\underset{}{}}{\overset{\overset{OH}{|}}{C}}HCH_2CH_3$$

b.
$$CH_3CH_2-\overset{\overset{O}{||}}{C}-NHCH_2CH_3 + H_2O \xrightarrow{H_3O^+} CH_3CH_2-\overset{\overset{O}{||}}{C}-OH + CH_3CH_2NH_2$$

c.
$$CH_3CH_2-\underset{\underset{}{}}{\overset{\overset{OH}{|}}{C}}HCH_2CH_3 \xrightarrow{\text{conc. } H_3O^+} CH_3CH=CHCH_2CH_3$$

d. $CH_3OH_2^+ + CH_3NH_2 \longrightarrow CH_3NH_3^+ + CH_3OH$

e.

$$\xrightarrow{KMnO_4}$$

f.
$$CH_3CH_2-\overset{\overset{O}{||}}{C}-OH + CH_3CH_2OH \xrightarrow{H_3O^+} CH_3CH_2-\overset{\overset{O}{||}}{C}-OCH_2CH_3 + H_2O$$

g.

$$+ H_2O \xrightarrow{H_3O^+}$$

3. Work with your team to classify each of the molecules below as a primary, secondary, or tertiary alcohol.

a.

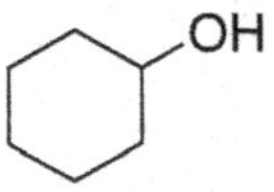

b.

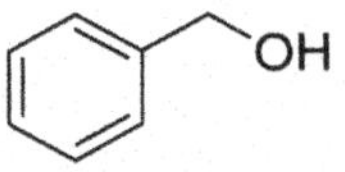

c. CH_3CH_2OH

d. 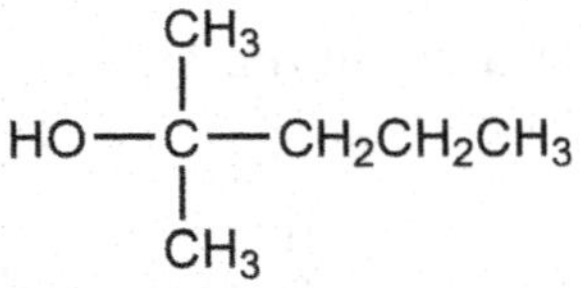
$$HO-\underset{\underset{CH_3}{|}}{\overset{\overset{CH_3}{|}}{C}}-CH_2CH_2CH_3$$

4. For each of the alcohols in CTQ 3, show the oxidation products, if any. If the oxidation takes place in two steps, show the product of each step. Check that all team members agree.

5. Expain how your team ensured that each member understood the material as you worked through the activity today.

6. Write any questions remaining with your team about the seven reaction types.

Exercises:

1. Each of the following carboxylic acids can donate a hydrogen atom to water. Complete a chemical reaction for each.

a. 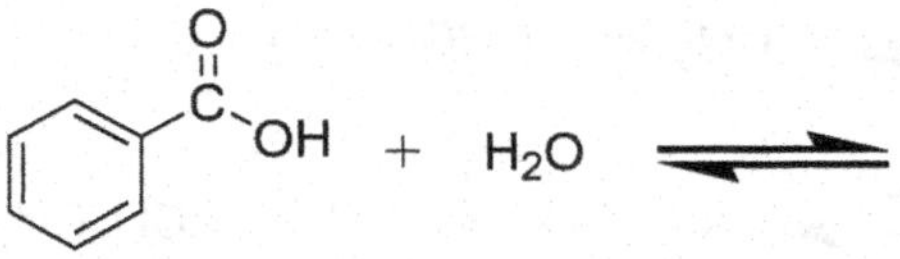 $+$ H_2O ⇌

b. acetic acid

c.

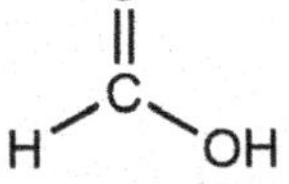

Information:

Just as it is beneficial to make a summary of the types of reactions, it is also useful to make a list of the functional groups that commonly react by each of the seven pathways. Such a summary is shown in Table 2.

Table 2: Functional groups that commonly react by the seven reaction pathways

acid-base	– carboxylic acids, phenols (acids)
	– amines (bases)
addition	– alkenes, alkynes
elimination	– alcohols
reduction	– alkenes, ketones
oxidation	– aldehydes, primary and secondary alcohols
condensation	– acid + alcohol, acid + amine
hydrolysis	– esters, amides (peptides)

Exercises:

2. Draw the reactants and products for an acid-base reaction between a carboxylic acid and amine of your choice.

3. Draw the reactants and products for the addition of water to cyclohexene.

4. The molecule shown below is used commercially as apricot flavoring.
    a. Name the molecule.
    b. Draw the carboxylic acid and alcohol that could combine in a condensation reaction to produce the molecule.

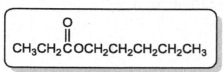

5. Read the assigned pages in your textbook and work the assigned problems.

Reactions of Alkanes and Alkenes
(What difference does a double bond make?)

Model 1: Reactions of alkanes

(1) $CH_4 + 2O_2 \longrightarrow CO_2 + 2H_2O$

(2) $CH_4 + Br_2 \xrightarrow{\text{heat or light}} CH_3Br + HBr$

Reaction 2 is a **halogenation reaction** and occurs with any halogen. The halogen substitutes for one of the hydrogens of the hydrocarbon.

Critical Thinking Questions:

1. Reaction (1) is what type of reaction? _____

2. Write a balanced reaction for C_4H_{10} that is the same type as reaction (1).

3. Write a balanced reaction of CH_4 with Cl_2.

4. Halogens can react more than once with a hydrocarbon. Take your hydrocarbon product from CTQ 3, and write a balanced halogenation reaction for it with Cl_2.

5. Work with your team to determine the maximum number of halogenation reactions CH_4 could undergo. Circle your team answer below.

1   2   3   4   5   6

Model 2: Alkene Reactions

(3) $\diagup C{=}C\diagdown + O_2 \longrightarrow CO_2 + H_2O$

(4) $\diagup C{=}C\diagdown + A{-}B \longrightarrow -\underset{A}{C}-\underset{B}{C}-$

Reaction (4) is an **addition reaction** in which atoms or groups of atoms are added to each carbon of a carbon-carbon double (or triple) bond.

Critical Thinking Questions:

6. What type of reaction is reaction (3)? _____

7. a. If C_2H_4 is reacted with O_2, what are the products?

    b. As a team, write and balance the reaction between C_2H_4 with O_2.

8. What happens to the double bond of the alkene in reaction (4)?

9. a. Write a reaction for C_2H_4 with the generic example A-B.

    b. How do the bond angles around C change from reactant to product?

    c. What happens to the hydrogens in the reaction? (Are they lost as in substitution reactions like reaction 2?) ***Recorder:*** write the consensus answer in a complete sentence.

Model 3: Types of addition reactions

Critical Thinking Questions:

10. In Model 3, there may be four different types of reactions, depending on the identity of A and B. Work with your team to match the name of the reaction for the listed reactant.

    a. H_2 i.   hydration

    b. Cl_2 ii. hydrogenation

    c. HOH   iii. hydrohalogenation

    d. HBr   iv. halogenation

11. a. As a team, predict the product that results when 2-pentene (five carbon chain with a double bond between the second and third carbons) is reacted with Br_2.

   b. Based on your answer to CTQ 10, what type of reaction is this?

12. Because reaction (6) is asymmetrical (H and B are not the same atom), two products are possible. One is given in Model 3. Predict the other possible product of the reaction. **Presenter:** Be prepared to share your team's consensus answer.

13. Experimentally, one product (the one in the model) predominates.
   a. What is different about the two possible products?

   b. As a team, determine how you would predict the major product formed in a reaction of type (6). Write your answer in a complete sentence.

Information:

The dominant product of hydrohalogenation and hydration reactions can be predicted following **Markovnikov's rule**. He observed that the H in the reaction will add to the carbon of the double bond with the greater number of H's. When the alkene is symmetrical, a single product results.

Critical Thinking Questions:

14. As a team, review the reactions for alkenes. Determine a way to remember all the reactions by only learning reactions (3) and (4).

15. As a team, discuss what difference the double bond makes. In other words, how do alkenes differ from alkanes in their reactivity? **Recorder:** Write the consensus answer in a complete sentence.

16. List any remaining questions about alkane and alkene reactions.

Exercises:

1. Write structures for all the possible products of the first halogenation reaction of C_3H_8 with Cl_2.

2. Write the organic products of the following reactions:

a.

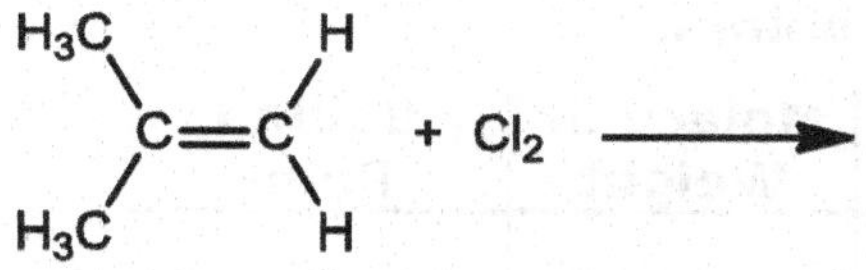

$$\begin{array}{c} H_3C \\ \\ H_3C \end{array} C=C \begin{array}{c} H \\ \\ H \end{array} \; + \; Cl_2 \longrightarrow$$

b.

$$\begin{array}{c} H_3C \\ \\ H_3C \end{array} C=C \begin{array}{c} H \\ \\ CH_3 \end{array} \; + \; HBr \longrightarrow$$

c.

$$\begin{array}{c} H \\ \\ H_3CH_2CH_2C \end{array} C=C \begin{array}{c} H \\ \\ H \end{array} \; + \; H_2 \longrightarrow$$

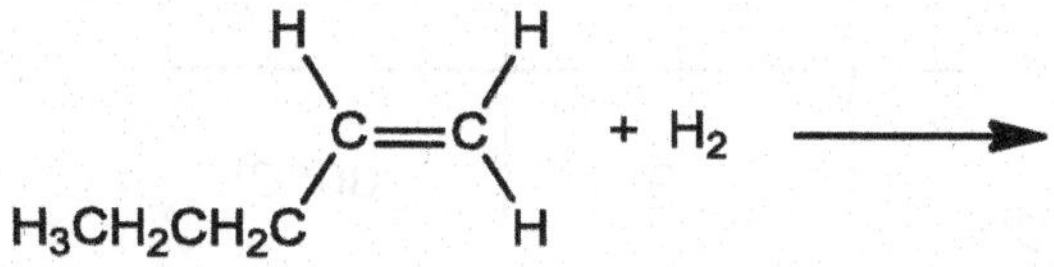

d. 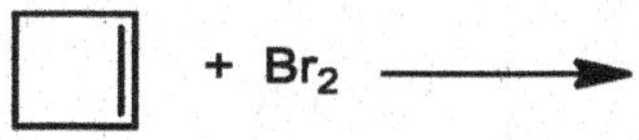 $\square\!\!\!|$ $+ \; Br_2 \longrightarrow$

e.

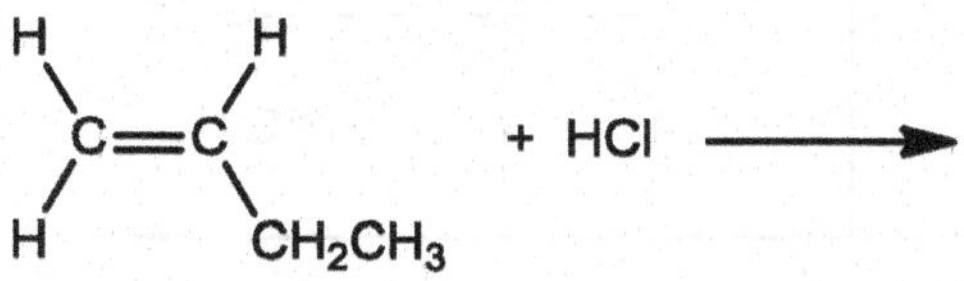

$$\begin{array}{c} H \\ \\ H \end{array} C=C \begin{array}{c} H \\ \\ CH_2CH_3 \end{array} \; + \; HCl \longrightarrow$$

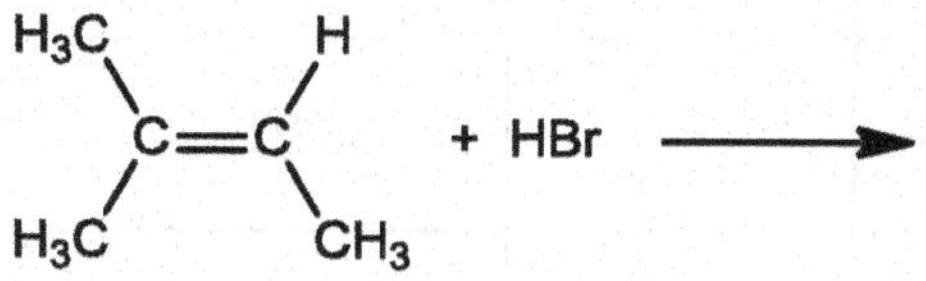

f.

$$\begin{array}{c} H_3C \\ \\ H_3CH_2C \end{array} C=C \begin{array}{c} H \\ \\ CH_3 \end{array} \; + \; HOH \longrightarrow$$

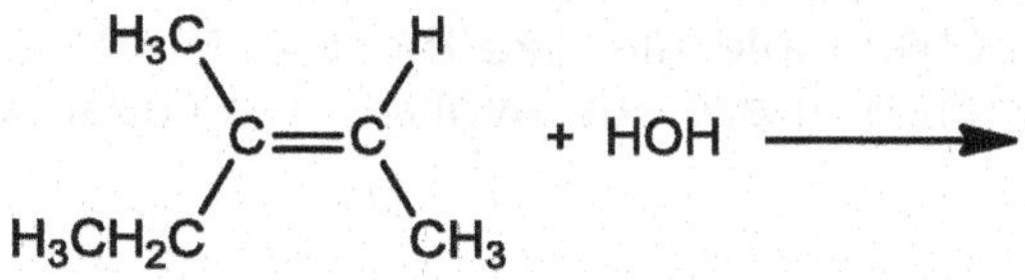

3. Read the assigned pages in your textbook and work the assigned problems.

Oxygenated Compounds
(What difference does an oxygen make?)

Model 1. Functional groups containing oxygen

Oxygen bonding options:

$$:\ddot{O}- \qquad :\ddot{O}=$$

Table 1. Organic functional groups and boiling points.[*]

Name	Example	Molecular Weight	Boiling Point
Alcohol	$H_3CH_2CH_2CH_2C-OH$	74	118°C
Ether	$H_3CH_2C-O-CH_2CH_3$	74	35°C
Aldehyde	$H_3CH_2CH_2C-\overset{\overset{\displaystyle O}{\|\|}}{C}-H$	72	75°C
Ketone	$H_3CH_2C-\overset{\overset{\displaystyle O}{\|\|}}{C}-CH_3$	72	80°C
Carboxylic Acid	$H_3CH_2C-\overset{\overset{\displaystyle O}{\|\|}}{C}-OH$	74	141°C
Ester	$H_3C-\overset{\overset{\displaystyle O}{\|\|}}{C}-O-CH_3$	74	57°C

Critical Thinking Questions:

1. Consider an oxygen atom contained in a neutral molecule, like those in Model 1. Give the name for the electrons around the oxygen that are not involved in a covalent bond.

2. What bond angle is formed when oxygen forms two single bonds? What shape is formed?

3. Which functional groups in Model 1 are polar? Compare with your team.

---

[*] Source: CRC Handbook of Chemistry and Physics, 92nd ed., 2011.

4. When a carbon atom forms a double bond to an oxygen atom, it is called a **carbonyl group**. Which functional groups in Model 1 contain a carbonyl group?

5. How does the structure of an alcohol differ from an ether?

6. Work with your team to describe how an aldehyde differs in structure from a ketone.

7. Thiols are compounds which resemble alcohols, except that the oxygen atom is replaced by a sulfur atom. Draw the analogous thiol for the four carbon alcohol in Table 1.

8. As a team, describe the structural difference between carboxylic acids and esters. Write your team answer in a complete sentence.

9. a. Are ethers polar molecules? _____

   b. Would you expect ethers to have **higher** or **lower** boiling points than alkanes (circle one)? Explain.

   c. Pentane (an alkane) has a boiling point of 36°C. Does the data agree with your prediction? Discuss with your team and explain why this could be the case.

10. Consider pentane again for the following questions.
    a. How does pentane's boiling point compare with the boiling points of the alcohol and the aldehyde in Model 1?

    b. Can molecular weight differences account for the observed differences in boiling points?

    c. Work with your team to propose an explanation for the observed differences in boiling points.

11. Compare the boiling points of the ketone and carboxylic acid. Work with your team to propose an explanation for the observed differences in boiling points.

12. Why do carboxylic acids have the highest boiling point, even higher than alcohols? Work with your team to explain this observation.

Model 2: Solubility

For solubility of organic molecules, a rule of thumb is that *"like dissolves like."* For example, polar substances are soluble in polar solvents, and non-polar in non-polar. However, the size of the molecule can affect solubility.

Figure 1. Solubility of carboxylic acids of increasing molecular weight.

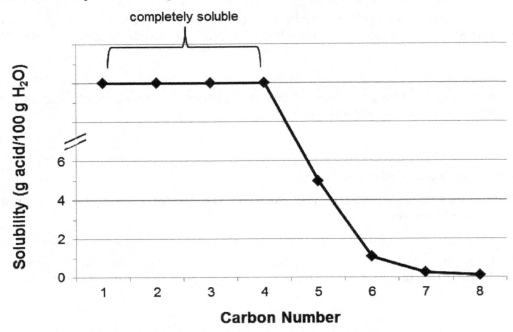

Critical Thinking Questions:

13. For each pair of small molecules below, predict which one would be more water soluble.
    a. An alcohol or an alkane? Why?

    b. An ether or an aldehyde? Why?

    c. A ketone or a carboxylic acid? Why?

14. Would propanoic acid (the carboxylic acid example in Table 1) be soluble in water? _____ Explain how you know this.

15. Work with your team to devise a consensus explanation for why solubility decreases with increasing carbon chain length.

16. Which functional group that we have studied so far has the highest boiling point? Explain why its boiling point is the highest.

17. Summarize the major point(s) of the activity.

18. What did you encounter today that is still confusing to one or more team members?

Information:

Forces of various strengths attract molecules to each other. Polar molecules are attracted to each other due to *dipole-dipole interactions*, which are simply attractions between the partial positive and negative charges in polar molecules. A particularly strong type of dipole-dipole interaction is found in molecules that contain hydrogen atoms that are bonded to oxygen or nitrogen atoms; this attraction is called *hydrogen bonding*. Even nonpolar molecules are slightly attracted to each other, because random variations in the electron clouds around the atoms cause tiny dipoles to appear and disappear continually; these are called *London dispersion forces*, and are proportional to the surface area (that is, the *size*) of the molecules.

Exercises:

1. Identify the functional group in each of the following molecules. Then give an acceptable name for each molecule.

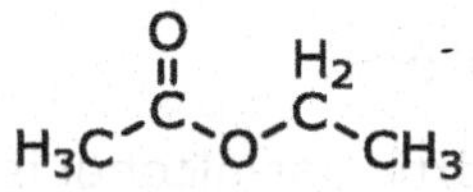

Ethers

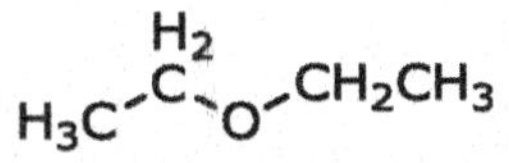

Ethers

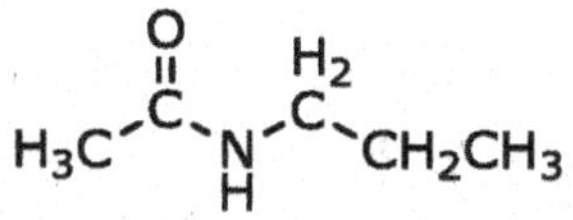

Amides

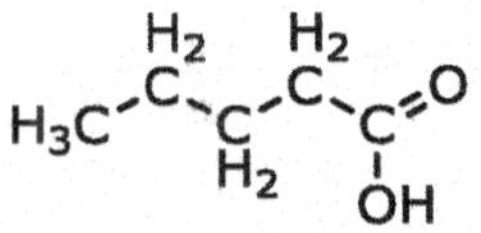

Carboxylic Acids

2. Which of the molecules in Exercise 1 will exhibit hydrogen bonding?

Amine and Carboxylic Acids

3. Predict which compound in each pair would have the higher boiling point. Which forces account for this difference?

a.

H₃C—CH₂—C(=O)—OH   soluble

H₃C—CH₂—CH₂—CH₂—CH₂—C(=O)—OH

4 or less

b.

H₃C—CH₂—CH(OH)—CH₂—CH₃   H₃C—CH₂—CH₂—CH₂—C(=O)—OH

c.

H₃C—C(=O)—O—CH₃   H₃C—CH₂—O—CH₂—CH₃   behave like alkanes

4. Examine the molecules in Exercise 3. Choose the term from the following list that describes the solubility of each molecule in water: soluble, slightly soluble, not soluble.

5. Predict the compound with the higher boiling point. Explain.
    a. butane or CH₃CH₂CH₂OH   Alcohols can form hydrogen bonds. Alkanes are non polar
    b. CH₃CH₂OCH₂CH₂CH₃ or CH₃CH₂CH₂COOH
    c.

    H₃CH₂C—C(=O)—H   or   H₃C—C(=O)—CH₃

6. Butanoic acid is found in butter and parmesan cheese. Explain why its boiling point is much higher (164°C) than ethyl acetate (77°C) despite their molecular weights being the same.

    H₃CH₂CH₂C—C(=O)—OH   H₃C—C(=O)—O—CH₂CH₃

    Butanoic acid   Ethyl acetate

    Carboxylic acid has both dipole-dipole and H bonding attractions so they need higher boiling points to break their bonds

7. Which of the following are soluble in water?
    a. CH₃CH₂CH₂OH   c. CH₃CH₂CH₂CH₃
    b. CH₃CH₂CH₂OCH₂CH₃   d. CH₃CH₂CH₂CH₂COOH

8. Review your textbook to learn how to determine whether a particular alcohol is primary, secondary, or tertiary.

9. Read the assigned pages in your textbook and work the assigned problems.

Reactions of Alcohols
(Does oxygen make a difference?)

Model 1: Preparation of alcohols

(1) 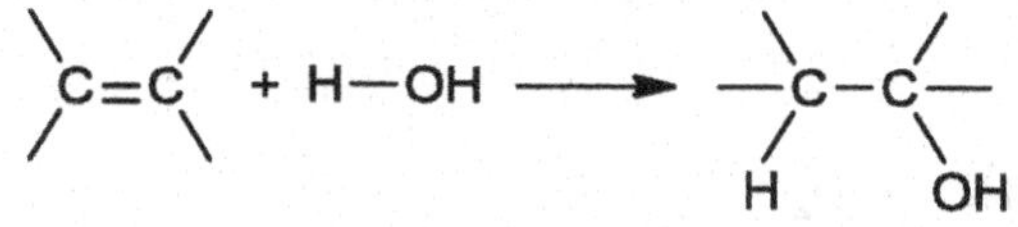 $\overset{\backslash}{\underset{/}{C}}=\overset{/}{\underset{\backslash}{C}}$ + H—OH $\longrightarrow$ $\overset{\backslash}{\underset{/}{C}}\underset{H}{\overset{|}{-}}\overset{/}{\underset{\backslash}{C}}\underset{OH}{\overset{|}{-}}$

(2) 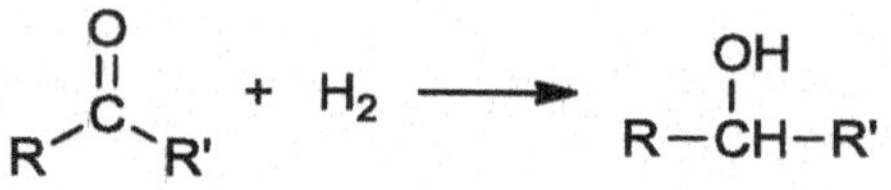 $\underset{R}{\overset{O}{\overset{\|}{C}}}\overset{}{R'}$ + H$_2$ $\longrightarrow$ $R-\underset{}{\overset{OH}{\underset{|}{CH}}}-R'$ (R and R' can be H or an organic group)

Critical Thinking Questions:

1. What is the name for the reaction type in (1)? _Addition_

2. In reaction (1), a hydrogen atom is added to one carbon of the C-C double bond, and an –OH group is added to the other carbon. Discuss with your team what happens to the double bond in reaction (2), and write a sentence to describe the reaction.

   An Alkene reacts with water and the double bond converts to a single bond.

3. Work with your team to determine what product is formed when water is added to 2-methyl-2-pentene.

4. Work with your team to determine what product is formed when hydrogen gas is added to 2-butanone (shown below).

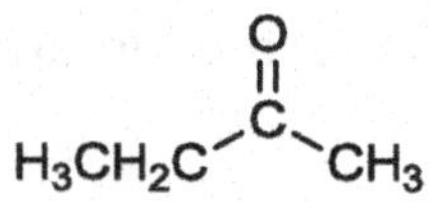

   $\underset{H_3CH_2C}{\overset{O}{\overset{\|}{C}}}\overset{}{CH_3}$

Model 2: Elimination (dehydration) reactions of alcohols

Elimination reactions require catalysts of sulfuric acid and high temperatures (180°C).

(3)

Example:

Critical Thinking Questions:

5. Elimination reactions are the reverse of what type of reactions? _____

6. Work with your team to describe what happens in reaction (3) in a sentence or two.

7. When reaction (3) has more than one way to eliminate water, different products may be possible. Work with your team to predict a possible product of the dehydration example reaction in Model 2 that has the double bond in a different location.

8. Experimentally, one product (the one in the model) predominates.

   a. What is different about the two possible products?

   b. Discuss as a team and propose a method to predict the product formed in an elimination reaction of an alcohol.

Information:

The dominant product of elimination reactions can be predicted following **Zaitsev's rule** which states the major product is the alkene with the greatest number of alkyl groups attached to the carbon atoms of the double bond.

Critical Thinking Questions:

9. Work with your team to predict the major product of the dehydration of:

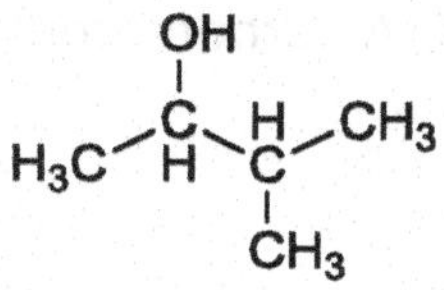

$$H_3C-\overset{\overset{\displaystyle OH}{|}}{\underset{H}{C}}-\overset{H}{\underset{\underset{\displaystyle CH_3}{|}}{C}}-CH_3$$

10. What alcohol can be dehydrated to produce 4-methyl-2-hexene?

Model 3: Acidic properties of alcohols

(4) $CH_3CH_2OH + H_2O \rightleftharpoons CH_3CH_3O^- + H_3O^+$

(5) 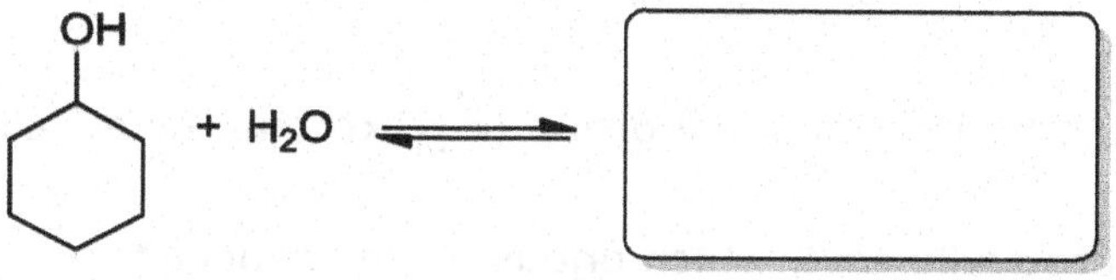 $+ H_2O \rightleftharpoons$ [ ]

Critical Thinking Questions:

11. a.   For reaction (4) in Model 3, what acts as the acid in the forward reaction?

   b.   What acts as the base in the forward reaction?

12. Complete reaction (5) above. Identify the acid and base for the forward reaction.

13. Complete following acid-base reaction:

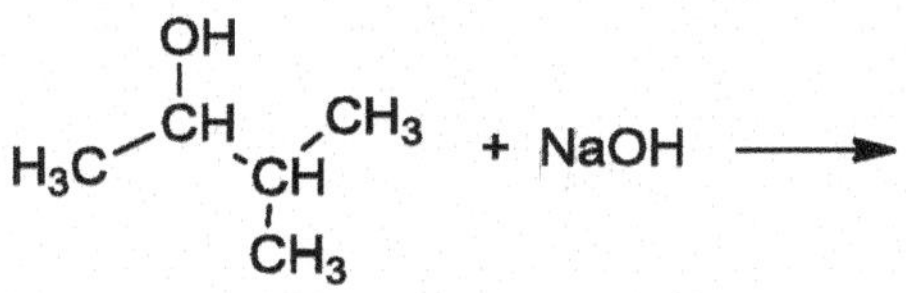

$$H_3C-\overset{\overset{\displaystyle OH}{|}}{CH}-\underset{\underset{\displaystyle CH_3}{|}}{CH}-CH_3 + NaOH \longrightarrow$$

Model 4: Oxidation and reduction of organic compounds

Organic redox reactions could be classified using oxidation numbers as with inorganic compounds; however, the task would be more difficult. To simplify matters, a carbon atom is considered *oxidized* if it loses hydrogen atoms or gains oxygen atoms. A carbon atom is considered *reduced* if it gains hydrogen atoms or loses oxygen atoms.

Oxidation:

(6) $R-\overset{\overset{\displaystyle OH}{|}}{\underset{\underset{\displaystyle H}{|}}{C}}-H \xrightarrow{[o]} R-\overset{\overset{\displaystyle O}{\|}}{C}-H \xrightarrow{[o]} R-\overset{\overset{\displaystyle O}{\|}}{C}-OH$

(7) $R-\overset{\overset{\displaystyle OH}{|}}{\underset{\underset{\displaystyle H}{|}}{C}}-R' \xrightarrow{[o]} R-\overset{\overset{\displaystyle O}{\|}}{C}-R' \xrightarrow{\;\times\;} R-\overset{\overset{\displaystyle O}{\|}}{C}-OH$

[o] represents the addition of an oxidizing agent.

Reduction:

Addition of H_2 as in Reaction (2) in Model 1 above.

Critical Thinking Questions:

14. For reaction (6), label each molecule by its functional group (*e.g.* ketone, ester, alcohol, *etc.*).

15. When comparing an alcohol to a carboxylic acid, which one is more reduced? _____

16. When comparing an alcohol to a ketone, which one is more oxidized? _____

17. What functional group is produced in the *reduction* of an aldehyde with H_2? _____

18. What is the difference between the reactants in reactions (6) and (7)?

19. a.   What product is produced from the oxidation of $CH_3CH_2CH_2OH$?

    b.   Can the product from part (a) be further oxidized? ____ If so, what is the new product?

20. a.   What product is produced from the oxidation of the following alcohol?

$$H_3C-\overset{\overset{\displaystyle OH}{|}}{\underset{}{C}}H-\overset{H_2}{\underset{H_2}{C}}-\overset{H_2}{C}-CH_3 \xrightarrow{[o]}$$

    b.   Can the product from part (a) be further oxidized? ____ If so, what is the new product?

21. Explain how one can predict the product for elimination reactions.

22. What questions remain with your team about alcohol reactions?

23. What insight have you gained today?

24. Did everyone contribute to the activity today? If so, explain how. If not, identify what you need to do to achieve equal participation by all next time.

Exercises:

1. Summarize the major reactions in this activity.

2. Predict the alcohol needed to produce the following product:

   a. $H_2C=CH_2$ b. 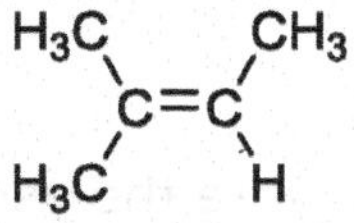 H_3C CH_3 c. 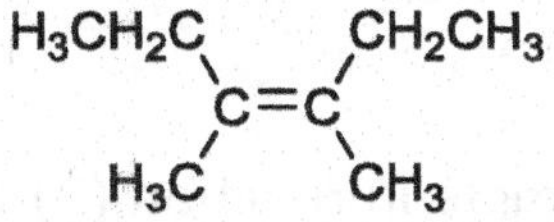 H_3CH_2C CH_2CH_3

   $C=C$ $C=C$

   H_3C H H_3C CH_3

   $H_2C - CH_3$

   $|$

   OH

   H_3C CH_3

   $H - C = C$

   H $|$ $|$ H

   OH H

   CH_3CH_2 H OH CH_2CH_3

   $C = C$

   $|$ CH_3

   H_3C

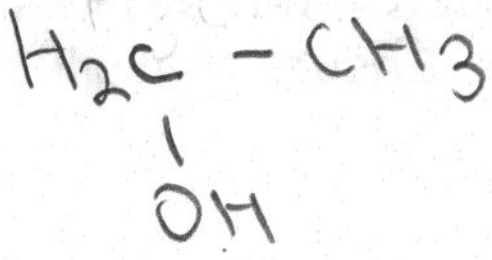

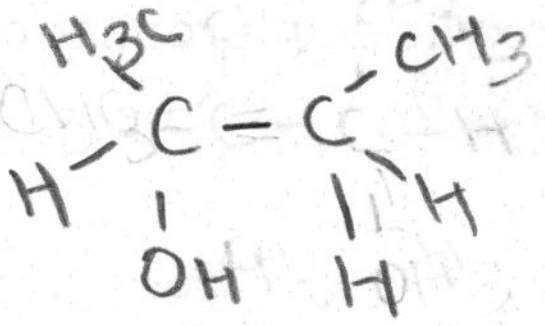

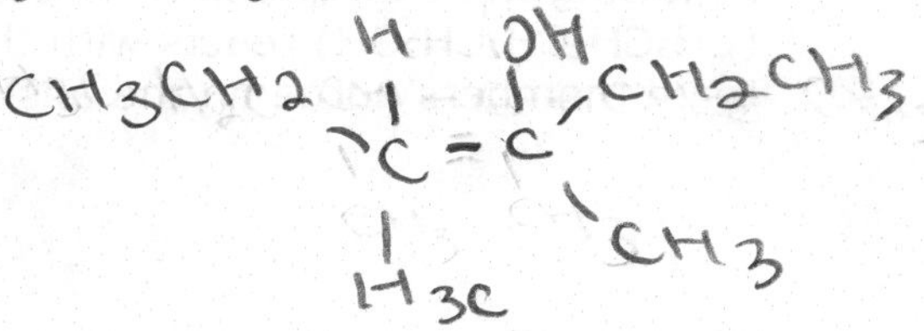

3. Draw the product when the following is oxidized:

   a. 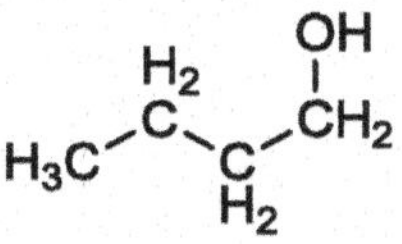   b. 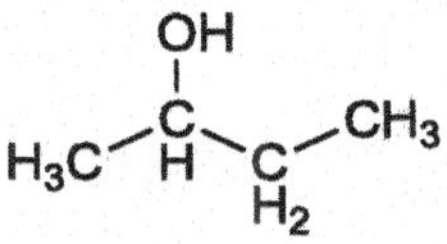   c.

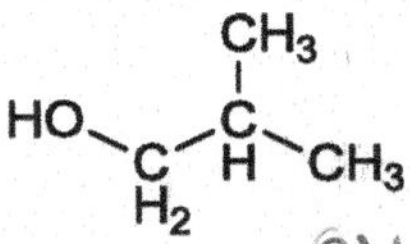

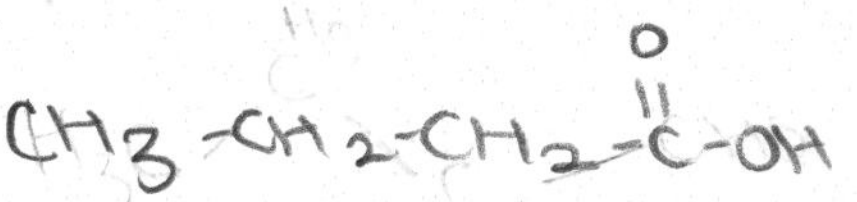

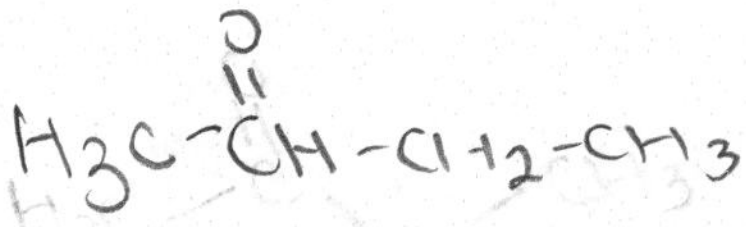

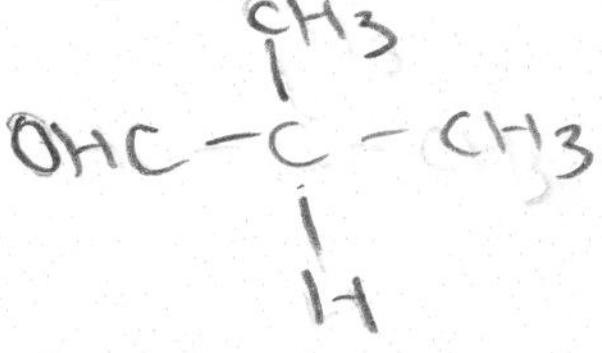

   OH

   H_2 $|$

   C CH_2

   H_3C C

   H_2

   OH

   $|$

   C CH_3

   H_3C H C

   H_2

   CH_3

   $|$

   HO C C CH_3

   H

   H_2

   $CH_3 - CH_2 - CH_2 - \overset{O}{\overset{||}{C}} - OH$

   $H_3C - \overset{O}{\overset{||}{C}}H - CH_2 - CH_3$

   CH_3

   $|$

   $OHC - C - CH_3$

   $|$

   H

4. Read the assigned pages in the text, and work the assigned problems.

- 203 -

CA40

Reactions of Carboxylic Acids and Esters
(What differences do two oxygens make?)

Model 1: Formation of carboxylic acids and esters

(1)

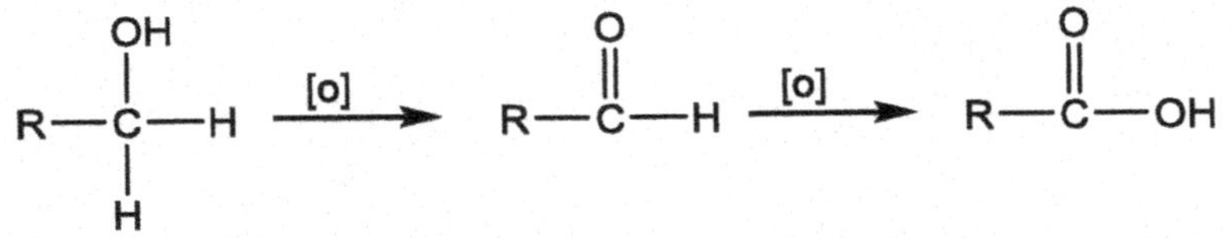

(2)

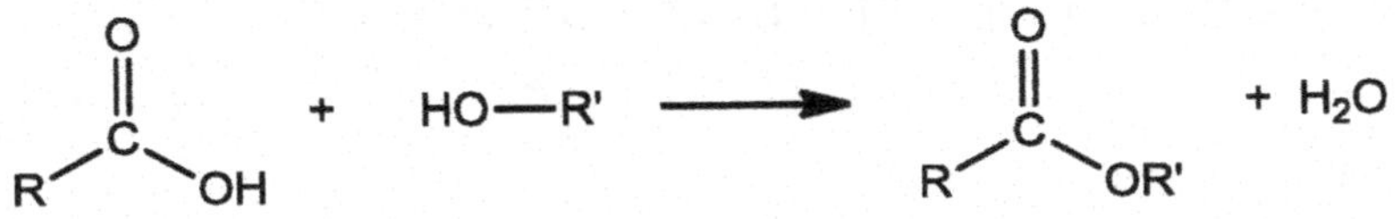

Critical Thinking Questions:

1. Which is more oxidized: an alcohol or a carboxylic acid? Explain.

2. Which is more reduced: a ketone or an alcohol? Explain.

3. Draw the carboxylic acid product formed from the complete oxidation of the following alcohol. Compare your answer with your team.

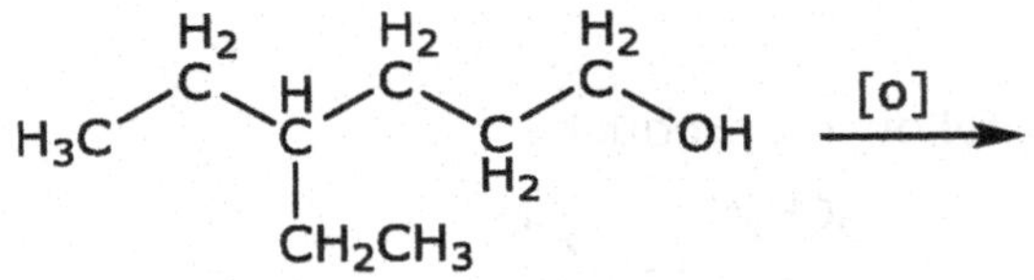

4. Following the example in reaction (2) of Model 1 write the ester formed when butanol ($CH_3CH_2CH_2CH_2OH$) reacts with the carboxylic acid product of CTQ 3. Ensure that all team members agree on the answer.

Model 2: Reactions of carboxylic acids and esters

(3)

$$\underset{R}{\overset{O}{\parallel}}\overset{\displaystyle C}{\underset{OH}{}} + H_2O \longrightarrow \underset{R}{\overset{O}{\parallel}}\overset{\displaystyle C}{\underset{O^{\ominus}}{}} + H_3O^{\oplus}$$

(4)

$$\underset{R}{\overset{O}{\parallel}}\overset{\displaystyle C}{\underset{OH}{}} + NaOH \longrightarrow \underset{R}{\overset{O}{\parallel}}\overset{\displaystyle C}{\underset{O^{\ominus}\,^{\oplus}Na}{}} + H_2O$$

(5)

$$\underset{R}{\overset{O}{\parallel}}\overset{\displaystyle C}{\underset{OR'}{}} + H_2O \longrightarrow \underset{R}{\overset{O}{\parallel}}\overset{\displaystyle C}{\underset{OH}{}} + R'\!\!-\!\!OH$$

(6)

$$\underset{R}{\overset{O}{\parallel}}\overset{\displaystyle C}{\underset{OR'}{}} + NaOH \longrightarrow \underset{R}{\overset{O}{\parallel}}\overset{\displaystyle C}{\underset{O^{\ominus}\,^{\oplus}Na}{}} + R'\!\!-\!\!OH$$

Critical Thinking Questions:

5. Which reactions in Model 2 are acid-base reactions?

6. Identify and label the conjugate base of the carboxylic acid in reaction (3).

7. When a carboxylic acid reacts with a strong base, a carboxylic acid salt forms with the cation of the base. Which reaction in Model 2 illustrates this? ____ Indicate the products of the following reaction. Does your team agree?

$$H_3CHCH_2C\overset{\overset{\textstyle O}{\parallel}}{\underset{\underset{\textstyle H_3C}{|}}{}}\!\!-\!\!OH + KOH \longrightarrow$$

8. Reaction (2) in Model 1 (the formation of an ester) can be reversed to give back the carboxylic acid and alcohol. What are the products of the following reaction? Compare with your team.

$$H_3CH_2CH_2CHC\overset{\overset{\textstyle O}{\parallel}}{\underset{\underset{\textstyle CH_3}{|}}{}}\!\!-\!\!O\!\!-\!\!CH_2CH_2CH_3 + H_2O \longrightarrow$$

Information:

Reaction (6) is the hydrolysis of an ester in basic solution, and is called **saponification**. This process has traditionally been used to make soap by adding lye (NaOH) to vegetable oils or animal fats to break the ester linkage.

Critical Thinking Questions:

9. Which reactions in Model 2 are hydrolysis reactions?

10. Write the soap formed in the following reaction.

OCH_3 + NaOH $\longrightarrow$

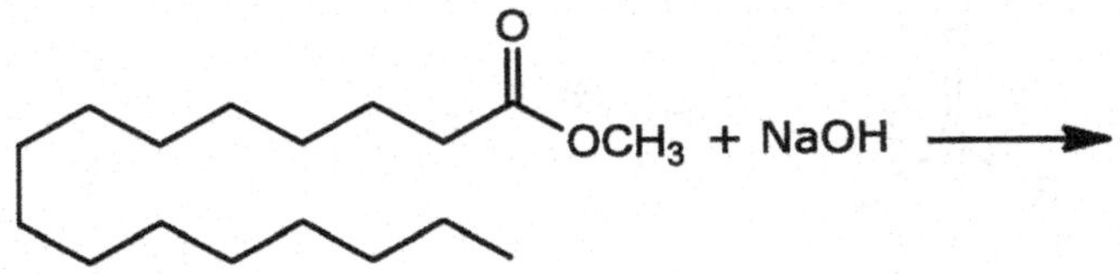

11. Summarize the major reactions in this activity. Are any of the reactions the reverse of other reactions in the activity? Are any similar to each other?

12. What question(s) remain about this activity?

Exercises:

1. Draw the carboxylic acids formed by the oxidation of each of the following molecules.

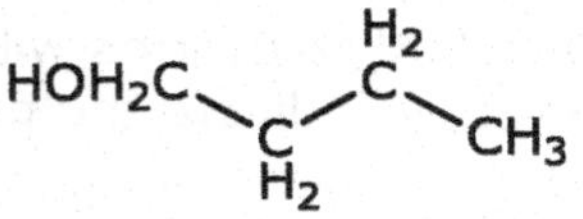

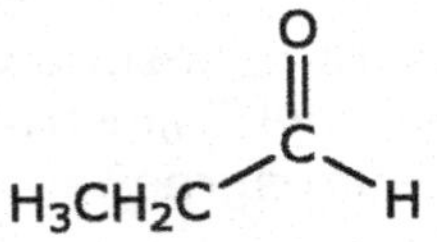

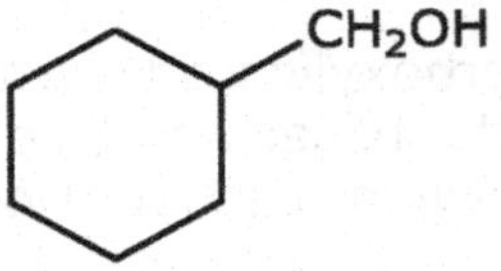

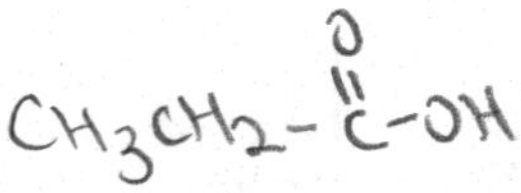

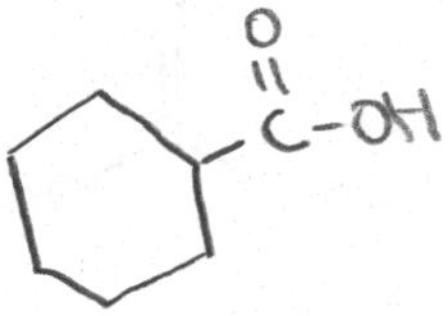

$$CH_3CH_2-\overset{O}{\underset{||}{C}}-OH$$

$$HO-\overset{O}{\underset{||}{C}}-CH_2-CH_2-CH_3$$

2. Draw the esters formed by the following reactions.

a.

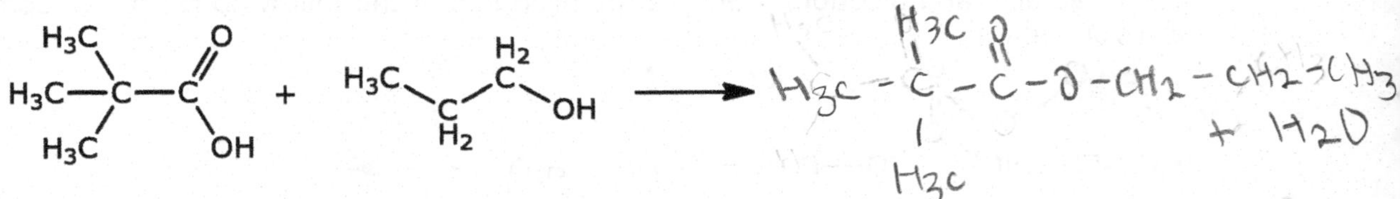

$$H_3C-\overset{CH_3}{\underset{|}{\overset{||}{C}}}-\overset{O}{\underset{||}{C}}-O-CH_2-CH_2-CH_3$$
$$+ H_2O$$

b.

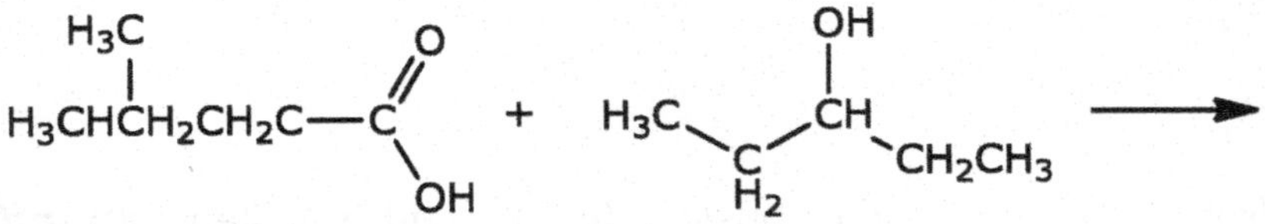

3. Indicate the products of the following reaction:

+ H_2O $\longrightarrow$ $CH_3CH_2CH_2$—$\overset{O}{\underset{||}{C}}$—$OH$ + CH_3OH

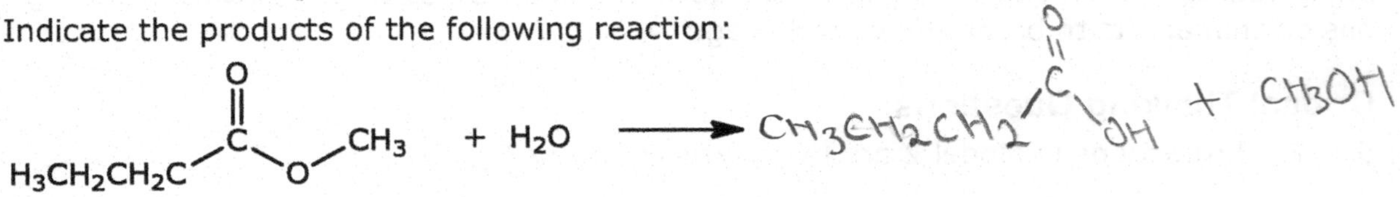

4. Read the assigned pages in your text, and work the assigned problems.

CA41   - 206 -

Properties of Amines and Amides
(What difference does nitrogen make?)

Model 1: Structures and boiling points of selected organic molecules

Nitrogen bonding options:

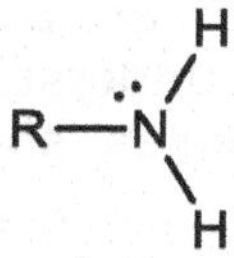

amine 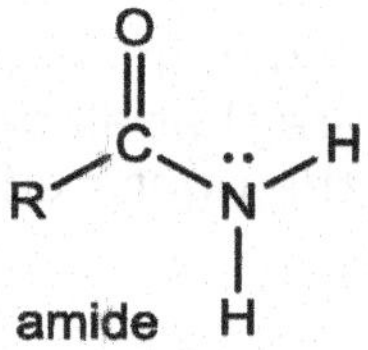   amide

Functional Group	Example Structure	Molecular Weight	Boiling Point
Alkane	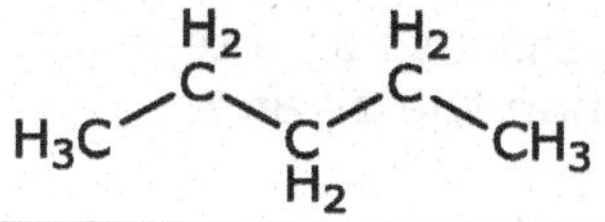	72	36°C
Amine (A)	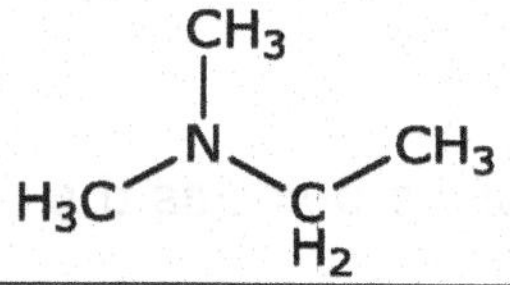	73	36°C
Amine (B)	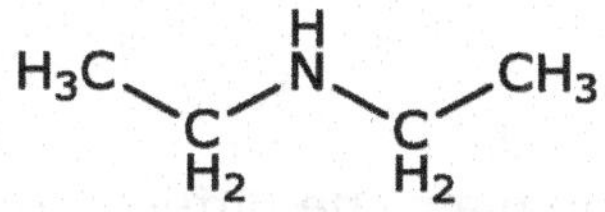	73	56°C
Amine (C)	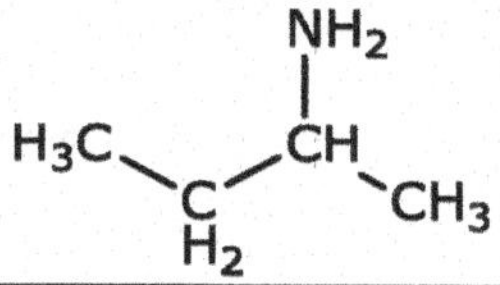	73	63°C
Amine (D)	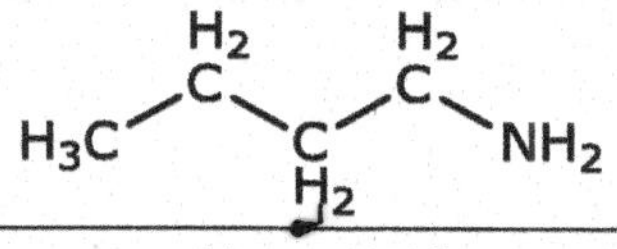	73	78°C
Alcohol	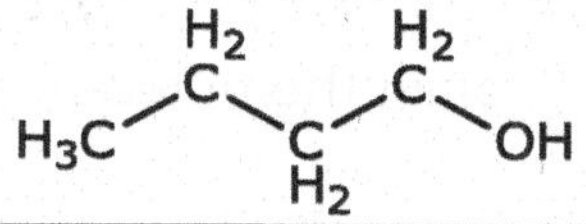	74	118°C
Carboxylic Acid	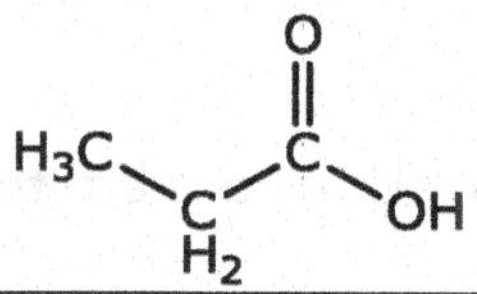	74	141°C
Amide	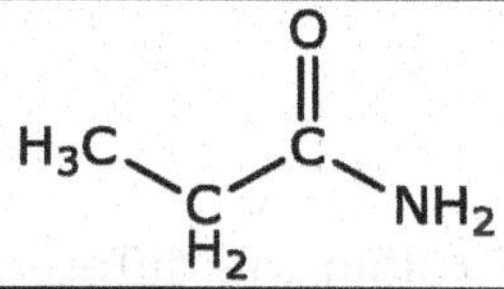	73	218°C

Critical Thinking Questions:

1. Label each Amine (A–D) in Table 1 as primary, secondary, or tertiary.

2. Which classes of amines – primary, secondary, or tertiary – can participate in hydrogen bonding with other identical molecules? Discuss as a team, and record your consensus.

3. How many different combinations of hydrogen bonds are possible between the following two molecules? ____ Ensure that all team members understand how to arrive at this answer. In the space at the right, redraw the molecules, showing a hydrogen bond between them.

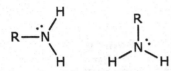

4. a. Compare Amines A–D in Model 1. Discuss as a team, and explain why three of the amines have a higher boiling point than the alkane.

   b. One of the amines has the same boiling point as the alkane. Explain this result.

5. Compare Amine (A) and Amine (B). Explain the difference in boiling point.

6. Compare Amine (B) with amine (C). Explain the difference in boiling point.

7. Compare Amine (C) with amine (D). Explain the difference in boiling point.

8. Compare Amine (D) with the alcohol. Work as a team to explain the difference in boiling point.

9. Compare the amide with the alcohol. Explain the difference in boiling point.

10. Would you expect amines and amides with short carbon chains to be soluble in water? Work as a team to explain your answer. (Hint: Recall the solubility of carboxylic acids discussed in CA 39.)

11. Which class of amine has the highest boiling point? Why?

12. What was one strength of your team today? Why was this a strength?

Exercises:

1. Rank the following amines from lowest to highest boiling point. Explain your reasoning.

2. Which compound in each pair would have the higher boiling point? Why?

   a.

   b.

3. Circle each of the following molecules that would be significantly soluble in water. Explain.

4. Read the assigned pages, and work the assigned problems.

Reactions of Amines and Amides
(What difference does a nitrogen make?)

Critical Thinking Question:

1. When a compound acts as a base, does it **donate** or **accept** a proton (H^+)? _____

Model 1: Acid-base reactions of amines

$$(1) \quad H_3C-\overset{\overset{H}{|}}{\underset{\underset{H}{|}}{N}}: + H_2O \rightleftharpoons H_3C-\overset{\overset{H}{|}}{\underset{\underset{H}{|}}{\overset{\oplus}{N}}}-H + \overset{\ominus}{O}H$$

$$(2) \quad H_3C-\overset{\overset{H}{|}}{\underset{\underset{H}{|}}{N}}: + HCl \rightleftharpoons H_3C-\overset{\overset{H}{|}}{\underset{\underset{H}{|}}{\overset{\oplus}{N}}}-H + \overset{\ominus}{C}l$$

Critical Thinking Questions:

2. Circle the base in the reactants for reaction (1). Repeat for reaction (2).

3. Explain why the products of the reactions in Model 1 have charges, even though the reactants do not. Compare your answer with your team.

Manager: Ask a different team member to offer the first explanation for CTQs 4–7.

4. What are the products of the reaction:

$$\underset{\underset{CH_3CHCH_2CH_2CH_3}{|}}{NH_2} + H_2O \rightleftharpoons$$

5. As a team, predict the organic product of the following reaction.

$$\underset{H_3CH_2C}{\overset{\overset{H}{|}}{\underset{}{N}}}\diagdown_{CH_3} + HBr \rightleftharpoons$$

6. Do amines typically react as acids or bases? _____

7. What are the products of the following reaction?

$$H_3CH_2CH_2C-NH_2 + CH_3COOH \rightleftharpoons$$

8. What is the organic product of the following:

$$\text{(cyclopentane)}-NH_2 + HCl \rightleftharpoons$$

9. Draw the products of the acid-base reaction.

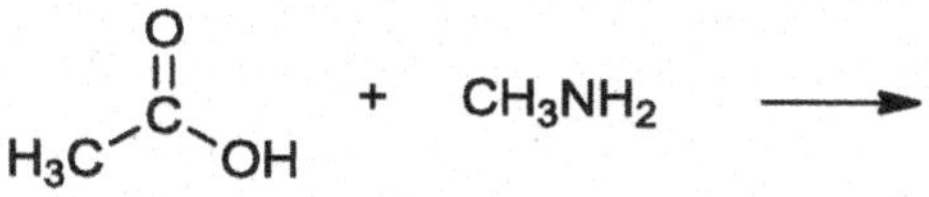

Model 2: Biochemical reactions of amides

In the laboratory, the reactants in Reaction (3) in Model 2 would normally undergo an acid-base reaction (as in CTQ 9). However, if the appropriate enzymes are present (for example, in a liver cell), the reaction can proceed as shown below.

Formation (_____)

(3)

Hydrolysis

(4)

Critical Thinking Questions:

10. Which of the seven classes of biochemical reactions (from CA 37) is reaction (3)? Write your team's answer into the blank in the parentheses after the word "Formation" in Model 2.

11. What are the two functional groups that react in reaction (3)?

12. Does reaction (3) form a **primary, secondary**, or **tertiary** amide? _____

13. Draw an arrow to the bond that is broken in the hydrolysis of the amide in reaction (4). What two functional groups are found in the products?

14. For the following hydrolysis reaction, is the reactant a **primary, secondary**, or **tertiary** amide (circle one)? As a team, predict the neutral products of the reaction.
    Manager: Ensure every member of the team understands before continuing.

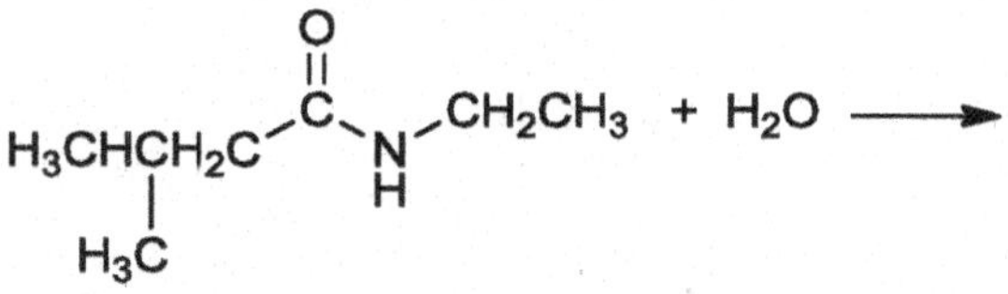

15. For the following condensation reaction:
    a. Circle the H and OH that are eliminated to become the H_2O in the products.
    b. Draw the structure of the other product into the box.
    c. Is the reactant amine **primary, secondary**, or **tertiary** (circle one)? Is the product a **primary, secondary**, or **tertiary** amide (circle one)? Does your team agree?

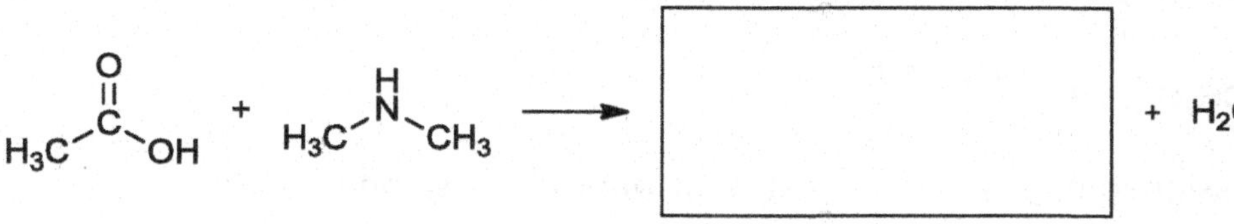

CA43

16. What are the **condensation** products of the following reaction? Does your team agree?

$$NH_3 \quad + \quad H_3C-\overset{\overset{\displaystyle O}{\|}}{C}-OH \quad \longrightarrow$$

17. What are the acid-base products, using the same reactants as in CTQ 16?

$$NH_3 \quad + \quad H_3C-\overset{\overset{\displaystyle O}{\|}}{C}-OH \quad \longrightarrow$$

18. The reactants below can undergo an acid-base reaction, but not a condensation reaction. Is the reactant amine **primary**, **secondary**, or **tertiary** (circle one)? Discuss with your team to determine why a condensation reaction is not possible in this case, and write your team explanation in a sentence.

$$H_3CH_2C-\overset{\overset{\displaystyle O}{\|}}{C}-OH \quad + \quad H_3C-\overset{\overset{\displaystyle CH_3}{|}}{N}-CH_3 \quad \longrightarrow$$

19. List any questions remaining with your team about reactions of amines and amides.

20. Give an example of how the manager's role today helped ensure that all team members understood the material.

Exercises:

Write the products of the following reactions of amines.

1.
$$H_3CH_2C-\overset{\overset{\displaystyle H}{|}}{N}-CH_2CH_3 \quad + H_2O \quad \rightleftharpoons$$

2.
$$H_3C-\overset{\overset{\displaystyle CH_3}{|}}{\underset{\underset{\displaystyle CH_2CH_3}{|}}{C}}-NH_2 \quad + H_2SO_4 \quad \rightleftharpoons$$

3.
(benzene ring)$-CH_2NH_2 \quad + \quad HCl \quad \rightleftharpoons$

4.
$$CH_3\overset{\overset{\displaystyle CH_3}{|}}{C}HCH_2\overset{\overset{\displaystyle O}{\|}}{C}-OH \quad + \quad H_3C-\overset{\overset{\displaystyle H}{|}}{N}-CH_3 \quad \longrightarrow$$
    (give both possible sets of products)

5.
$$H_3CH_2CH_2C-\overset{\overset{\displaystyle O}{\|}}{C}-\overset{\underset{\displaystyle H}{|}}{N}-CH_3 \quad + H_2O \quad \longrightarrow$$

6. Read the assigned pages in your text, and work the assigned problems.

Overview of Carbohydrates
(What makes a sugar?)

Model 1: Glyceraldehyde

Simple sugars have a molecular formula of $C_n(H_2O)_n$, meaning that for n carbon atoms in the sugar, there are n oxygen atoms and $2n$ hydrogen atoms. The simplest sugar (a monosaccharide) is glyceraldehyde, with the molecular formula $C_3H_6O_3$ and the structure shown below.

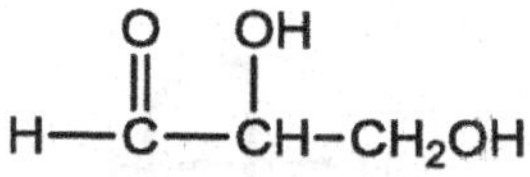

$$H-\overset{\overset{\displaystyle O}{\|}}{C}-\overset{\overset{\displaystyle OH}{|}}{CH}-CH_2OH$$

Critical Thinking Questions:

1. Circle the aldehyde functional group in the structure of glyceraldehyde above.

2. What is the value of n for glyceraldehyde in the formula $C_n(H_2O)_n$?

3. Hypothesize on the origin of the term *carbohydrate*.

4. Using a molecular model kit, make a molecule of glyceraldehyde. (In many model kits, black = C, red = O, white = H. Use short bonds for single bonds and <u>two</u> of the longer, flexible bonds for double bonds.)

    Hold your model with the aldehyde carbon at the top and the CH_2OH at the bottom, and arrange it so that the middle carbon is closer to you than the other two carbons.

    Now, compare your model with the two structures below. Circle the one that matches your model. (Remember what the dashes and wedges mean?)

5. Now make <u>a second</u> model of glyceraldehyde, so that you have one of each of the two structures shown above. Can you rotate or twist them so that all the atoms are in the same place in both molecules? _____

Model 2: Types of stereoisomers

Stereoisomers are molecules with the same connectivity, but different arrangements in space. This category includes geometric (*cis-trans*) isomers and a type of isomers we have not seen before called *enantiomers*. Enantiomers are a pair of non-identical molecules that are mirror images of each other.

Whenever a carbon is bonded to **four dissimilar groups**, it is said to be *chiral*. An object or molecule that is chiral is not identical to its mirror image. The term chiral literally means "handedness."

Critical Thinking Question:

6. Confirm that the center carbon of either glyceraldehyde in CTQ 4 is connected to four <u>different</u> (or dissimilar) groups.

7. Are the two forms of glyceraldehyde **geometric isomers** or **enantiomers** (circle one)? Explain in one complete sentence.

Model 3: L and D-glyceraldehyde

To distinguish the two isomers of glyceraldehyde in CTQ 4, chemists call the one on the left L-glyceraldehyde and the one on the right D-glyceraldehyde. One way to remember which is which is to imagine a line from C1 to C3 through the middle OH. The D isomer makes the letter "D." Nearly all common sugars of biological significance have the D configuration.

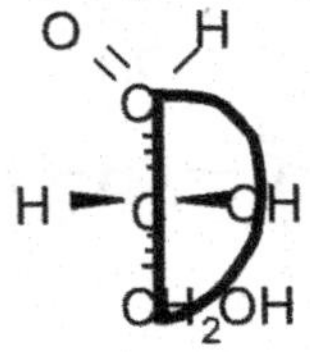

Figure 1: Wedge-and-dash structures and Fisher projections of glyceraldehyde

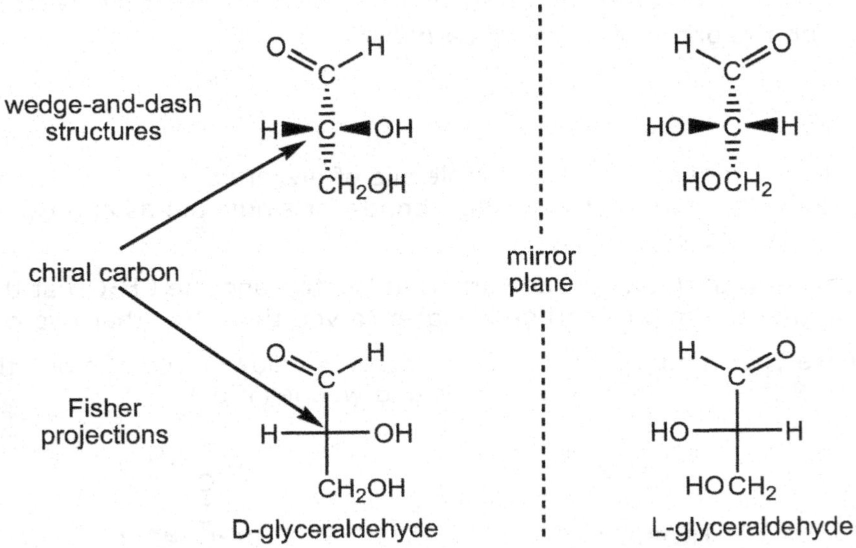

Critical Thinking Question:

8. The Fisher projection of L-glyceraldehyde shown below does not exactly match the Fisher projection in Figure 1. Explain how the two structures are different, and why both are correctly named. You may wish to look at and manipulate your model.

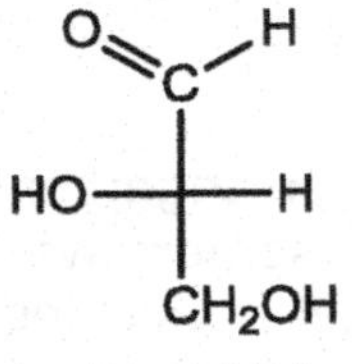

L-glyceraldehyde

Information:

The center carbon of glyceraldehyde is connected to four dissimilar (different) groups. This makes it a **chiral** carbon and leads to the two enantiomers. Each additional chiral carbon doubles the number of pairs of enantiomers, so that in 6-carbon carbohydrates with four chiral carbons, there are 16 possible isomers ($2^4 = 16$). In the Fisher projection, the last (bottom) chiral carbon determines whether the isomer is L or D.

Definitions:

aldose   –   a simple sugar containing an aldehyde
ketose   –   a simple sugar containing a ketone

Most of the common sugars have 5 or 6 carbons, and so are called **pentoses** or **hexoses**. You can mix the terms with **aldose** and **ketose** to get names such as **ketopentose** and **aldohexose**.

Figure 2: Common simple sugars

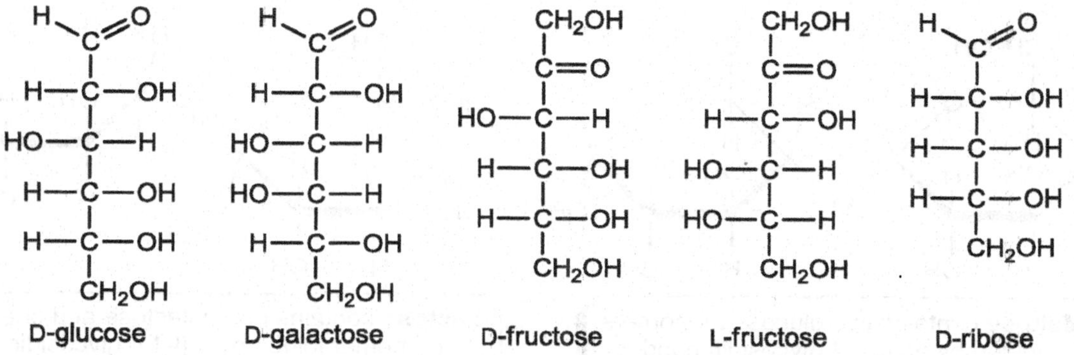

D-glucose   D-galactose   D-fructose   L-fructose   D-ribose

Critical Thinking Questions:

9. Label each sugar in Figure 2 with the appropriate name from this list: aldohexose, aldopentose, ketohexose, ketopentose.

10. Circle *each* chiral carbon in Figure 2. (Hint: There are 17 total).

11. Consider the two isomers of fructose in Figure 2. What is the difference between the L and D isomer of a simple sugar?

Information:

The simple 5- and 6-carbon sugars often react with themselves to make cyclic structures. When this happens, a new functional group is formed—a *hemiacetal*. A hemiacetal has an alcohol (-OH) and an ether (-OR) <u>attached to the same carbon</u>. An example is shown in Figure 3.

Figure 3: Cyclization of D-glucose to make α-D-glucose.

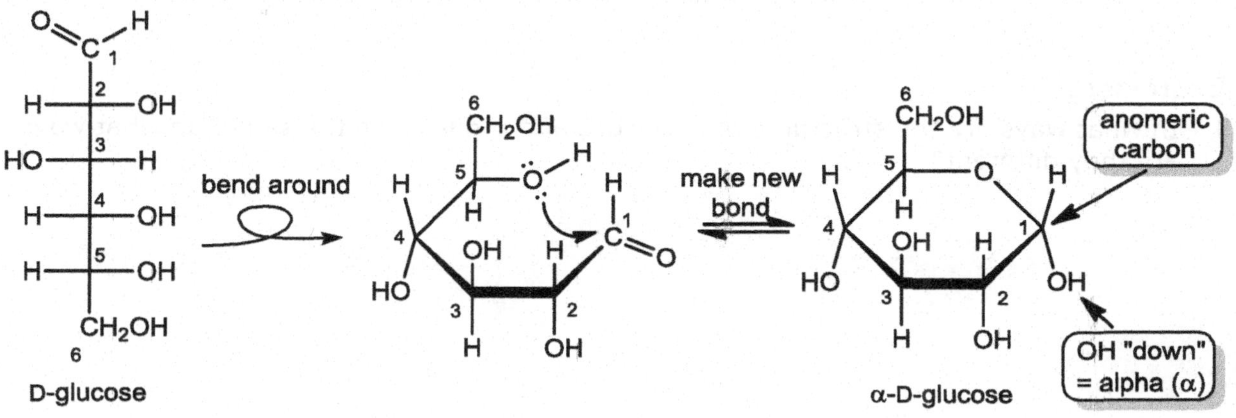

D-glucose   bend around   make new bond   α-D-glucose

anomeric carbon

OH "down" = alpha (α)

When this happens, a new chiral carbon is created. This carbon (the **anomeric** carbon) is designated α or β depending on whether its OH is "down" or "up" when the cyclic structure is drawn in the orientation shown in Figure 3. In disaccharides and polysaccharides, simple sugars are connected together with at least one bond at the anomeric carbon. Bonds to the anomeric carbon are called **glycosidic bonds**.

Figure 4: α and β glycosidic bonds in two disaccharides. (a) maltose, or malt sugar, and (b) lactose, or milk sugar.

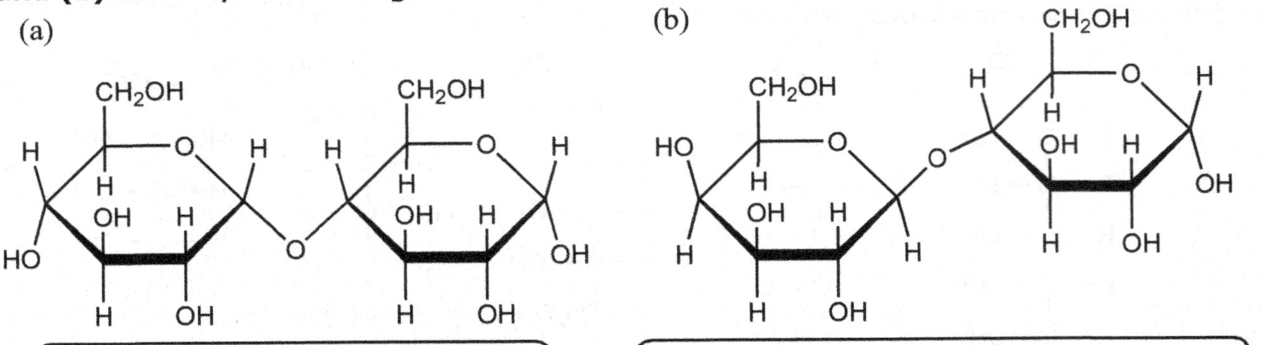

(a)

Maltose contains two glucose monomers joined by an α-1,4 glycosidic bond

(b)

Lactose contains one galactose and one glucose monomer joined by a β-1,4 glycosidic bond

Critical Thinking Questions:

12. Consider Figures 3 and 4. Explain what is meant by "α," "1," and "4" in the term α-1,4 glycosidic bond.

13. Explain what is meant by "β," "1," and "4" in the term β-1,4 glycosidic bond.

14. By extension, explain what would be meant by the term α-1,6 glycosidic bond.

Exercises:

1. In what ways are the structures of D-glucose and D-galactose the same? In what ways are they different?

2. Considering your answer to Exercise 1 and Figures 3 and 4, explain how you can determine which part of the disaccharide lactose in Figure 4 came from the glucose monomer, and which came from galactose. Then label the Figure accordingly.

3. Find descriptions in your textbook of the structures of the polysaccharides amylose, amylopectin, glycogen, and cellulose. Describe the similarities and differences among these polymers. Be sure to consider the types of glycosidic bonds as examined in CTQs 12-14.

4. Read the assigned pages in your textbook and work the assigned problems.

CA44A

Carbohydrate Structure
(What makes a simple sugar?)

Model 1: Structures of selected monosaccharides (simple sugars)

D-glucose   D-fructose   D-ribose

Critical Thinking Questions:

Manager: For CTQs 1–5, identify a different team member to give the first explanation.

1. Simple sugars have a molecular formula of $C_n(H_2O)_n$. What is the value of n for D-glucose in the formula $C_n(H_2O)_n$?

2. Hypothesize on the origin of the term *carbohydrate*.

3. Circle and label three different functional groups in the monosaccharides in Model 1.

4. Simple sugars are given the ending *–ose*. This can be added to the unique functional group of each to give a descriptive name, like aldose (for an aldehyde) or ketose (for a ketone). Identify the sugars in Model 1 as either an aldose or a ketose.

5. Monosaccharides can be further distinguished by the number of carbons present, for example: aldo*pent*ose, keto*tri*ose, etc. Add the appropriate name to the structures in Model 1. Compare your answers with your team.

Manager: Ensure that all team members understand classifying monosaccharides before continuing.

Model 2: Types of stereoisomers

Stereoisomers are molecules with the same connectivity, but different arrangements in space. This category includes geometric (*cis-trans*) isomers and a type of isomers we have not seen before called *enantiomers*. Enantiomers are a pair of non-identical molecules that are mirror images of each other.

Whenever a carbon is bonded to **four dissimilar groups**, it is said to be *chiral*. An object or molecule that is chiral is not identical to its mirror image. The term chiral literally means "handedness."

Critical Thinking Questions:

6. Using a molecular model kit, make a molecule of glyceraldehydes shown below. (In many model kits, black = C, red = O, white = H. Use short bonds for single bonds and <u>two</u> of the longer, flexible bonds for double bonds.)

   Hold your model with the aldehyde carbon at the top and the CH₂OH at the bottom, and arrange it so that the middle carbon is closer to you than the other two carbons.

   Now, compare your model with the two structures below. Circle the one that matches your model. (Remember what the dashes and wedges mean?)

7. Now make <u>a second</u> model of glyceraldehyde, so that you have one of each of the two structures shown above. Can you rotate or twist them so that all the atoms are in the same place in both molecules? _____

8. Confirm that the center carbon of either glyceraldehyde in CTQ 6 is connected to four <u>different</u> (or dissimilar) groups.

9. Are the two forms of glyceraldehyde **geometric isomers** or **enantiomers** (circle one)? Explain in one complete sentence.

Model 3: Wedge-and-dash structures and Fisher projections of glyceraldehyde

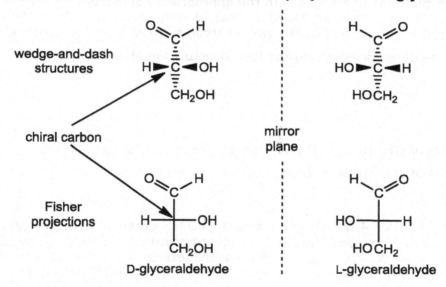

Critical Thinking Questions:

10. The Fisher projection of L-glyceraldehyde shown below does not exactly match the Fisher projection in Model 3. Explain how the two structures are different, and why both are correctly named. You may wish to look at and manipulate your model.

L-glyceraldehyde

CA44B

11. Identify the following as chiral or not chiral. **Manager:** *Ensure all team members agree.*

12. Consider the two isomers of glyceraldehyde in Model 3. What is the relationship between the L and D isomer of a simple sugar?

Model 4: Structures of selected monosaccharides

Critical Thinking Questions:

Manager: *For CTQs 13–16, identify a different team member to give the first explanation.*

13. Label each molecule in Model 4 with the appropriate name from this list: *aldopentose, aldotetrose, ketopentose, ketotetrose*. Then identify each molecule as the D or L isomer. (Note: D or L is determined by the chiral carbon farthest from the carbonyl carbon).

14. Describe the differences among the four structures in Model 4.

15. Which pairs of structures in Model 4 are mirror images of each other? _____

16. Which pairs of structures in Model 4 are not mirror images of each other? _____

Information:

All structures in Model 4 are **stereoisomers** of each other. More specifically, if two chiral molecules are mirror images, they are called **enantiomers**. If they are stereoisomers that are not mirror images, they are called **diastereomers**. A compound with n chiral carbon atoms has a maximum of 2^n possible stereoisomers, and half that many pairs of enantiomers.

Critical Thinking Questions:

17. What is the isomeric relationship between molecules A and C in Model 4? _____

18. What is the isomeric relationship between molecules B and C in Model 4? _____

19. Work with your team to identify all pairs of enantiomers and diastereomers in Model 4.

20. Write the difference between a diastereomer and an enantiomer. **Manager:** *Ensure all team members understand.*

21. For an aldopentose, how many stereoisomers are possible? _____ Enantiomers? _____

22. For a ketopentose, how many stereoisomers are possible? _____ Enantiomers? _____

23. List the different ways a carbohydrate can be classified.

24. List one area for improvement of your teamwork and how your team might accomplish that.

25. Cite an example of how the manager's role today helped ensure that everyone understood the material.

Exercises:

1. Classify the following monosaccharides as an aldose or ketose (including the number of carbons, such as *aldopentose, ketohexose, etc.*) and as the D or L isomer.

2. Classify the following pairs of structures as enantiomers, diastereomers, or neither.

3. Read the assigned pages in your textbook and work the assigned problems.

Carbohydrate Reactions
(How do carbohydrates react?)

Model 1: Ring formation with monosaccharides

The simple 5- and 6-carbon sugars often react with themselves to make cyclic structures. When this happens, a new functional group is formed—a *hemiacetal*. A hemiacetal has an alcohol (-OH) and an ether (-OR) <u>attached to the same carbon</u>. An example is shown in Figure 1.

Figure 1: Cyclization of D-glucose to make α-D-glucose.

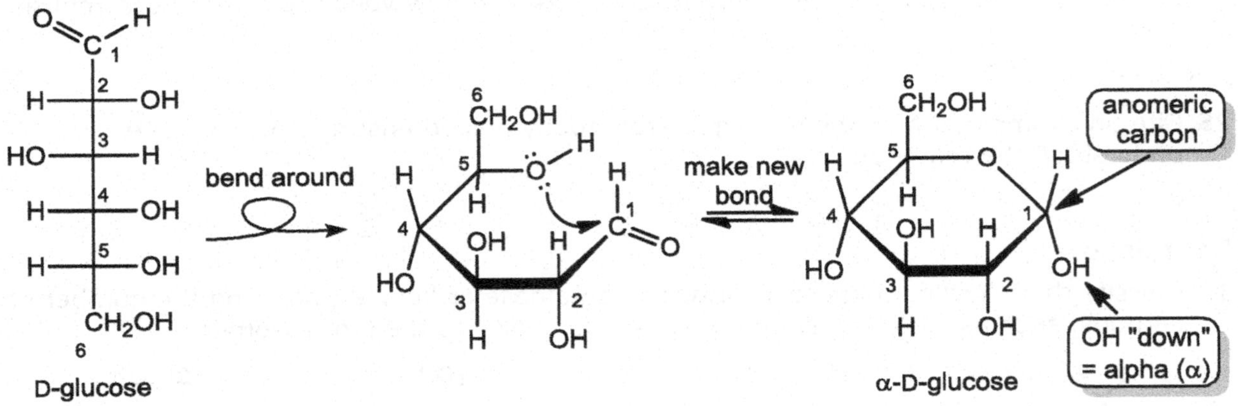

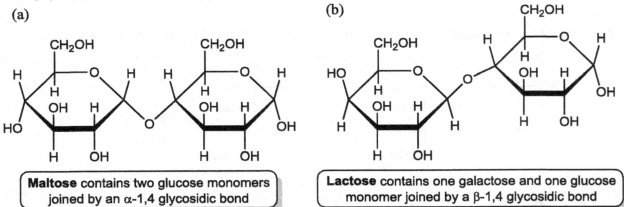

When this happens, a new chiral carbon is created. This carbon (the **anomeric** carbon) is designated α or β, depending on whether its –OH is "down" or "up" when the cyclic structure is drawn in the orientation shown in Figure 1. In disaccharides and polysaccharides, simple sugars are connected together with at least one bond at the anomeric carbon. Bonds to the anomeric carbon are called **glycosidic bonds**.

Figure 2: α and β glycosidic bonds in two disaccharides. (a) maltose, or malt sugar, and (b) lactose, or milk sugar.

(a)

(b)

Maltose contains two glucose monomers joined by an α-1,4 glycosidic bond

Lactose contains one galactose and one glucose monomer joined by a β-1,4 glycosidic bond

Critical Thinking Questions:

1. Circle the hemiacetal in α-D-glucose in Figure 1.

2. Which two carbons of glucose become connected *via* the hemiacetal when the ring forms?

3. Consider Figures 1 and 2. Explain what is meant by "α," "1," and "4" in the term α-1,4 glycosidic bond. Does your team agree?

4. Explain what is meant by "β," "1," and "4" in the term β-1,4 glycosidic bond.

5. By extension, explain what would be meant by the term α-1,6 glycosidic bond.

6. Consider the disaccharides in Figure 2. Would any of the monosaccharides be able to open back up into a straight chain in one step (as in Figure 1)? If so, which one(s)? Does your team agree? (Hint: consider the carbons involved in ring formation)

Model 2: Aldoses are reducing sugars.

D-glucose

Critical Thinking Questions:

7. When an aldehyde is oxidized, what functional group is produced?

8. Work with your team to draw the product for the reaction in Model 2 if only the aldehyde carbon is oxidized.

9. If the aldehyde in a carbohydrate is oxidized, how might this affect its ability to form a hemiacetal to close into a ring structure? Discuss with your team and write your consensus answer.

10. Review your answer to CTQ 6 and the structures for maltose and lactose in Figure 2. Would either of these disaccharides be able to function as a reducing sugar? Does your team agree?

11. In sucrose (table sugar), shown at the right, the anomeric carbons of glucose and fructose are connected together. Identify the linkage between the two monosaccharides, indicating α or β for both sugars. Could sucrose function as a reducing sugar?

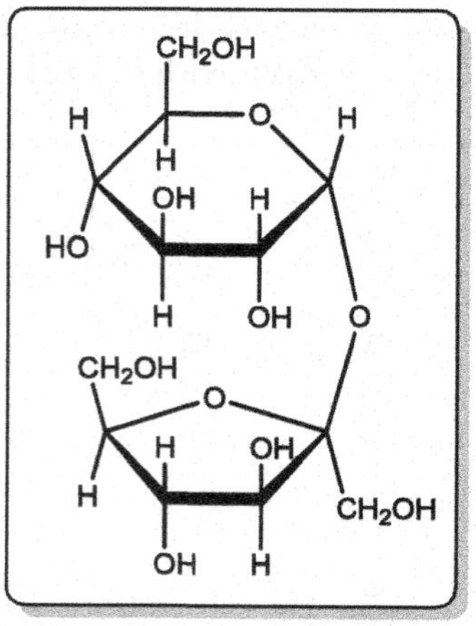

12. Work with your team to list the major concepts of this activity.

Exercises:

1. Classify the following monosaccharides as the α or β anomer.

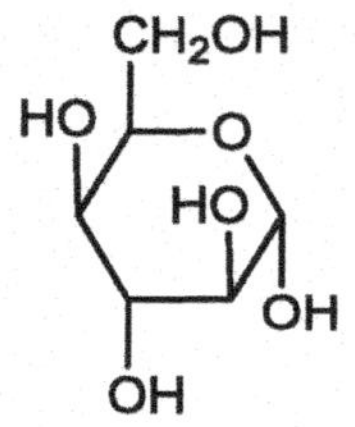

A-Anomer 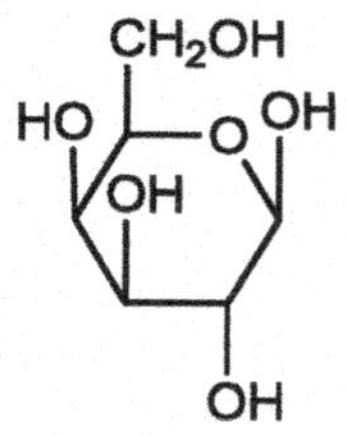   B-anomer 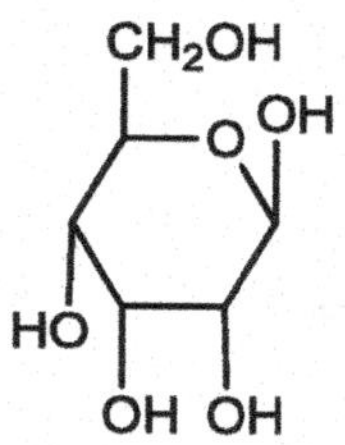   B-Anomer 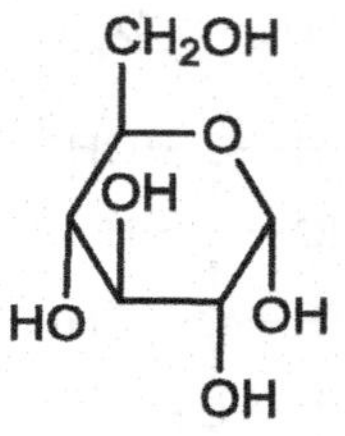   A-Anomer

2. Write the product when the following monosaccharide is oxidized.

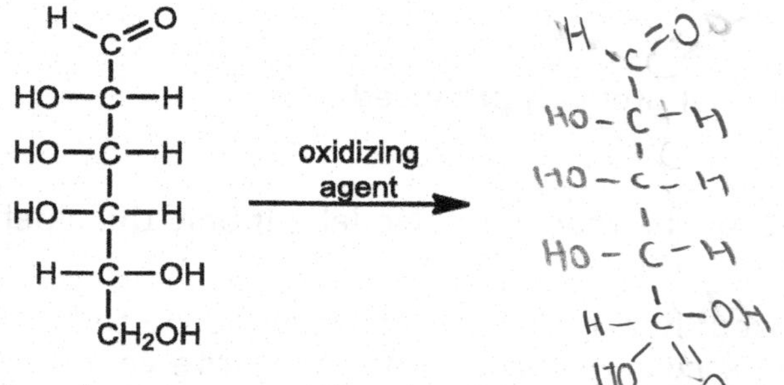

3. In what ways are the structures of D-glucose and D-galactose the same? In what ways are they different?

Both have Six Carbons, Same weight And formula.

The difference is the OH group on carbon 4.

4. Considering your answer to Exercise 3 and Figures 1 and 2, explain how you can determine which part of the disaccharide lactose in Figure 2 came from the glucose monomer, and which came from galactose. Then label Figure 2 accordingly.

5. Find descriptions in your textbook of disaccharides. Describe the similarities and differences among these. Be sure to consider the types of glycosidic bonds as examined in CTQs 3-5.

6. Find descriptions in your textbook of the structures of the polysaccharides amylose, amylopectin, glycogen, and cellulose. Describe the similarities and differences among these polymers. Be sure to consider the types of glycosidic bonds as examined in CTQs 3-5.

7. Read the assigned pages in your textbook and work the assigned problems.

CA44C

(This page is intentionally left blank.)

Overview of Lipids
(What are the components of cell membranes?)

Model 1: Structure of a fatty acid (palmitoleic acid) and its conjugate base

Fatty acids have a carboxyl group "head" and a hydrocarbon "tail" that is 11-19 carbons in length.

Critical Thinking Questions:

1. Circle and label the head and the tail of palmitoleic acid in Model 1.

2. Which would be more water soluble: palmitoleic acid, or sodium palmitoleate? Discuss with your team and write a consensus explanation.

3. a. Select any **one carbon** in Model 1 that is in a C=C bond, and circle it.
   b. One atom that is bonded to the carbon that you circled is not shown. What is the atom? _____ (You may draw it onto the diagram if you like.)
   c. What is the **total number** of atoms to which this carbon is covalently bonded? _____
   d. Does the double bond have the *cis* or *trans* configuration (circle one)?

Information:

A carbon is **saturated** when it is bonded to 4 atoms. A carbon in a carbon-carbon double bond (C=C) is bonded to only 3 atoms, and is said to be **unsaturated**. Fatty acids are often **unsaturated** because they contain one or more *cis* carbon-carbon double bonds. The double bond is referred to as a **site of unsaturation**.

Critical Thinking Questions:

Refer to Model 2 (on the following page) to answer CTQs 4-6.

4. Give a name of a fatty acid from Model 2 that fits the class listed.
   a. a saturated fatty acid _____
   b. a monounsaturated fatty acid _____
   c. a polyunsaturated fatty acid _____

5. Which class of fatty acid listed in CTQ 4 would be a solid at room temperature? _____

6. As a team, identify at least three fatty acids from Model 2 that have the same number of carbons (and roughly the same molar mass). **Circle** their names and melting points.
   a. How does the melting point change as the number of sites of unsaturation increases?

   b. Based on your answer to part (a), how do the intermolecular attractions between molecules change as the number of sites of unsaturation increases?

   c. Given the shapes of the molecular structures shown in Model 2, propose an explanation for the effect that you described in part (b) above.

Model 2: Structures and melting points of some common fatty acids[*]

Alkane	Number of carbons	Structure	Molar mass, g/mol	Melting point, °C
palmitic acid	16		256	62
stearic acid	18		284	69
arachidic acid	20		313	75
palmitoleic acid	16		254	0
oleic acid	18		282	13
linoleic acid	18		280	-5
linolenic acid	18		278	-11
arachidonic acid	20		304	-49

Monounsaturated fatty acids have one site of unsaturation.
Polyunsaturated fatty acids have more than one site of unsaturation.

Model 3: Composition of a generic fat molecule (triacylglygerol)

Chemically, **fats and oils** are **triacylglycerols** (triglycerides). They are composed of a glycerol "backbone" esterified with three fatty acids.

[*] Source: ChemFinder.com [accessed June 2006]

Critical Thinking Question:

7. Model 3 states that glycerol and fatty acids are "esterified" to make a triacylglycerol molecule. What class of organic reaction (from ChemActivity 37) is this esterification?

8. The main difference between a fat and an oil is that oils are liquids at room temperature.

   Would you expect the fatty acyl groups in an oil to be **more saturated** or

   more unsaturated than those in a fat? Circle one, and explain your choice.

Information: Cell membranes

The main lipid component of membranes is **glycerophospholipids**. The structure of a glycerophospholipid is like a fat molecule (triacylglycerol), but with one fatty-acyl "tail" replaced with a phosphate group plus an amino alcohol. This gives the glycerophospholipid one very polar "head" group with **two** nonpolar "tail" groups.

Figure 1: A phosphatidylethanolamine, a typical glycerophospholipid

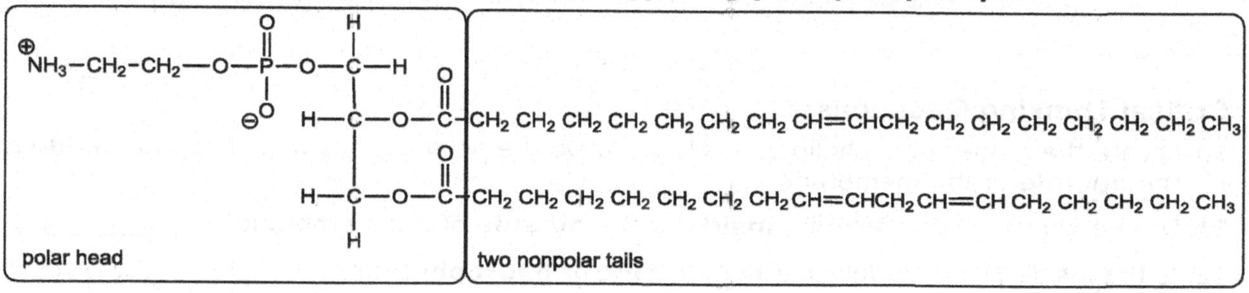

$\overset{\oplus}{NH_3}-CH_2-CH_2-O-\overset{\overset{O}{\|}}{P}-O-\overset{\overset{H}{|}}{\underset{|}{C}}-H$

$\overset{\ominus}{O}$ $H-\overset{|}{\underset{|}{C}}-O-\overset{\overset{O}{\|}}{C}-CH_2\ CH_2\ CH_2\ CH_2\ CH_2\ CH_2\ CH_2\ CH=CHCH_2\ CH_2\ CH_2\ CH_2\ CH_2\ CH_2\ CH_2\ CH_3$

$H-\overset{|}{\underset{|}{C}}-O-\overset{\overset{O}{\|}}{C}-CH_2\ CH_2\ CH_2\ CH_2\ CH_2\ CH_2\ CH_2CH=CHCH_2\ CH=CH\ CH_2\ CH_2\ CH_2\ CH_2\ CH_3$

H

polar head   two nonpolar tails

Critical Thinking Question:

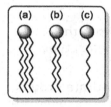
(a) (b) (c)

9. Of the three cartoons in the box at the left — (a), (b) or (c) — circle the one that would best represent a glycerophospholipid. Discuss as a team, and explain your choice.

Information:

Cell membranes also contain the steroid cholesterol. Because of the four fused rings in the structure, cholesterol is very rigid, and therefore adds rigidity to the membrane.

Figure 2: Structure of cholesterol as commonly drawn (a), and in the typical "rigid" chair conformation (b)

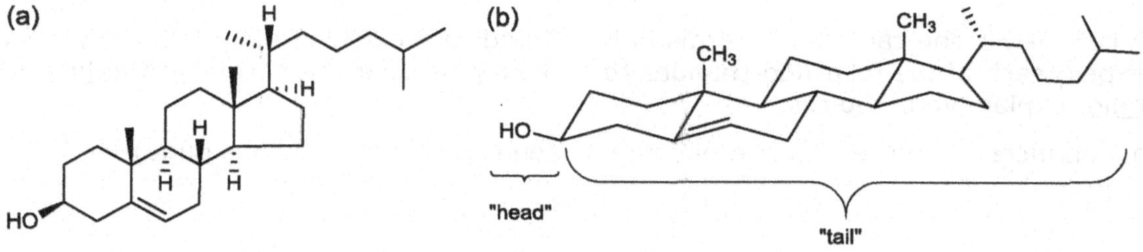

(a)

HO

(b)

CH₃

HO

"head"

"tail"

Model 4: Cross section of a typical cell membrane composed of a lipid bilayer with associated proteins.

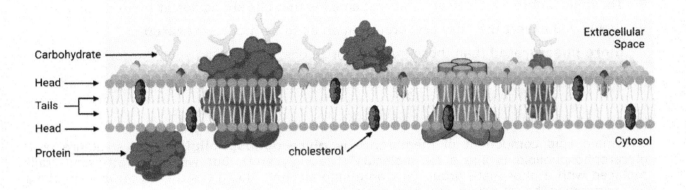

Created with BioRender.com

Critical Thinking Questions:

10. Locate the glycerophospholipids in Model 4. Do the polar head groups face the **inside** or the **outside** of the membrane? _____.

11. Do the nonpolar tails face the **inside** or the **outside** of the membrane? _____.

12. Is the inside of the membrane **hydrophilic** or **hydrophobic**? _____. Discuss with your team and explain why this is so.

13. Two types of proteins are found in cell membranes: integral (embedded into the membrane) and peripheral (on one side of the membrane). Label one of each type in Model 4.

14. Membrane proteins are often glycoproteins, that is, proteins with carbohydrate groups covalently attached to them. Locate and label two glycoproteins in Model 4. Explain why you would not expect any carbohydrate groups to be attached to the portion of the protein that is crossing the lipid bilayer.

15. Most globular proteins contain amino acids with polar side chains. Would you expect these proteins to be able to pass through a cell membrane? Explain why or why not.

16. A cell membrane can vary in terms of how "fluid" or flexible it is. Describe if you would expect each of the following changes to make a particular membrane <u>more fluid</u> or <u>more rigid</u>. Explain your choices.

   a. an increase in the cholesterol content (see Figure 2 on previous page)

b. an increase in the fraction of <u>unsaturated</u> fatty-acyl tails in the membrane glycerophospholipids

Information: Other types of lipids

Prostaglandins are formed from the 20-carbon fatty acid arachidonic acid. Various prostaglandins act like hormones in the body. Steroids and pharmaceuticals such as nonsteroidal anti-inflammatory drugs (NSAIDs) inhibit the formation of prostaglandins responsible for producing inflammation and pain.

Waxes are modified fatty acids in which the hydrogen of the carboxyl head is replaced with a second hydrocarbon tail. Many plants and animals produce or secrete waxes as a protective, water-repellent barrier.

Figure 3: Carnauba wax can be isolated from palm trees.

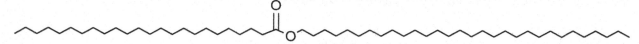

Exercises:

1. What organic functional group (other than alkane) is contained in a wax molecule?

2. Explain why a leaf coated with wax makes the leaf water-repellent.

3. Describe the differences between triacylglycerols and glycerophospholipids.

4. A common ingredient in many prepared foods is *partially hydrogenated soybean oil*. "Partially hydrogenated" means that some (but not all) of the double bonds in the fatty acid tails have been converted to single bonds. In this process, some of the remaining *cis* double bonds are also converted to *trans* double bonds.

    a. What would be the purpose of converting some of the double bonds in the fatty acid tails to single bonds? (How would this affect the properties of the oil?)

    b. How can you reduce the amount of *trans* fats in your diet?

5. Locate the section in your textbook that describes methods of transport of molecules across membranes. Using your own paper, describe the similarities and differences between simple *diffusion*, *active transport* and *facilitated transport*.

6. Read the assigned pages in your text, and work the assigned problems.

CA45A

Lipid Structure
(How are lipids different?)

Model 1: Structure of a fatty acid (palmitoleic acid)

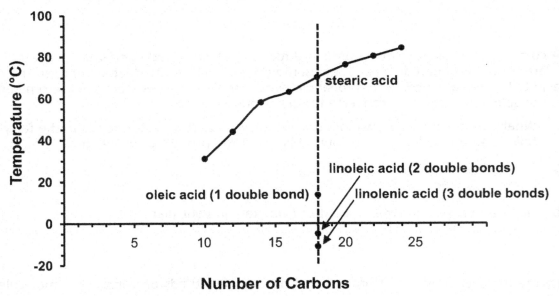

Fatty acids have a carboxyl group "head" and a hydrocarbon "tail" that is 11-19 carbons in length.

Critical Thinking Questions:

1. Circle and label the head and the tail of palmitoleic acid in Model 1.

2. What do the zig-zag lines in Figure 1 represent?

3. What functional groups are present in palmitoleic acid?

4. a. Select a carbon in Figure 1 that is in a C=C bond, and circle it. What is the **total number of atoms** to which this carbon is covalently bonded? _____

   b. Does the double bond have the *cis* or *trans* configuration (circle one)?

   c. Predict how a double bond like the one in Figure 1 would affect a fatty acid's ability to form a relatively straight chain.

Model 2: Melting points of selected fatty acids

Information:

A carbon is **saturated** when it is bonded to 4 atoms. A carbon in a carbon-carbon double bond (C=C) is bonded to only 3 atoms, and is said to be **unsaturated**. Fatty acids are often **unsaturated** because they contain one or more *cis* carbon-carbon double bonds. The double bond is referred to as a **site of unsaturation**.

Critical Thinking Questions:

5. How many carbons does stearic acid contain? _____

6. Identify stearic acid and linoleic acid as either saturated or unsaturated.

7. What is a melting point?

8. a. What happens to the melting point as the number of carbons in lipid molecules increases?

    b. As a team, write a consensus explanation for your observation in part (a).

9. a. What is similar between stearic acid and linoleic acid?

    b. What is different between stearic acid and linoleic acid?

    c. How does the difference between the two affect the melting point?

    d. Discuss as a team and propose a reason for this difference.

10. Write a team consensus for the two factors that affect the melting point of a lipid.

11. List any questions remaining with your team about fatty acid structure.

Exercises:

1. A common ingredient in many prepared foods is *partially hydrogenated soybean oil*. "Partially hydrogenated" means that some (but not all) of the double bonds in the fatty acid tails have been converted to single bonds. In this process, some of the remaining *cis* double bonds are also converted to *trans* double bonds.

   a. What would be the purpose of converting some of the double bonds in the fatty acid tails to single bonds? (How would this affect the properties of the oil?)

   b. How can you reduce the amount of *trans* fats in your diet?

2. Rank the following fatty acids in order from lowest to highest melting point.

   (a) palmitic acid (16 carbons, saturated fatty acid)

   (b) palmitoleic acid (16 carbons, one cis double bond)

   (c) arachidic acid (20 carbons, saturated fatty acid)

3. Read the assigned pages in your text, and work the assigned problems.

Lipids and Membranes
(What are the components of cell membranes?)

Model 1: Glycerphospholipids

The main lipid component of membranes is **glycerophospholipids**. The structure of a glycerophospholipid is like a fat molecule (triacylglycerol), but with one fatty-acyl "tail" replaced with a phosphate group plus an amino alcohol. This gives the glycerophospholipid one very polar "head" group with **two** nonpolar "tail" groups.

Figure 1: A phosphatidylethanolamine, a typical glycerophospholipid

$$\overset{\oplus}{N}H_3-CH_2-CH_2-O-\overset{\overset{O}{\|}}{\underset{\underset{\ominus}{O}}{P}}-O-\overset{\overset{H}{|}}{\underset{|}{C}}-H$$

$$H-\overset{|}{\underset{|}{C}}-O-\overset{\overset{O}{\|}}{C}\ CH_2\ CH_2\ CH_2\ CH_2\ CH_2\ CH_2\ CH_2\text{-}CH{=}CHCH_2\ CH_2\ CH_2\ CH_2\ CH_2\ CH_2\ CH_2\ CH_3$$

$$H-\overset{|}{\underset{\underset{H}{|}}{C}}-O-\overset{\overset{O}{\|}}{C}\ CH_2\ CH_2\ CH_2\ CH_2\ CH_2\ CH_2\ CH_2CH{=}CHCH_2\ CH{=}CH\ CH_2\ CH_2\ CH_2\ CH_2\ CH_3$$

Critical Thinking Question:

1. Identify and circle the polar head and tail regions of the glycerophospholipid in Figure 1.

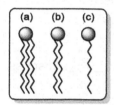

2. Of the three cartoons in the box at the left — (a), (b) or (c) — circle the one that would best represent a glycerophospholipid. Discuss as a team, and explain your choice.

3. Of the three cartoons in the box above — (a), (b) or (c) — circle the one that would best represent a fatty acid (CA45B). How does a glycerophospholipid differ from a fatty acid?

Model 2: Steroids

The other main lipid component of membranes is the steroid **cholesterol** (Figure 2). Because of the four fused rings in the structure, cholesterol is very rigid, and therefore adds rigidity to the membrane. Other steroids with the same ring structure are hormones such as estrogen and testosterone.

Figure 2: Structure of cholesterol as commonly drawn (a), and in the typical "rigid" chair conformation (b)

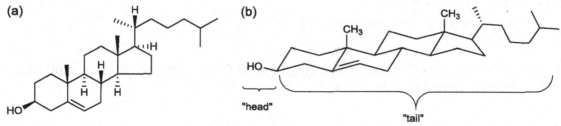

Model 3: Cross section of a typical cell membrane composed of a lipid bilayer with associated proteins.

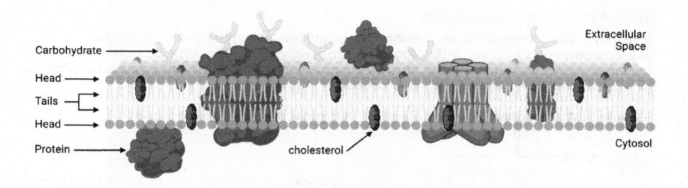

Created with BioRender.com

Critical Thinking Questions:

4. Work with your team to propose a reason why cells form lipid bilayers instead of monolayers (or micelles).

5. Is the inside of the membrane **hydrophilic** or **hydrophobic**? _____. Discuss with your team and explain.

6. Two types of proteins are found in cell membranes: integral (embedded into the membrane) and peripheral (on one side of the membrane). Label one of each type in Model 3.

7. Membrane proteins are often glycoproteins, that is, proteins with carbohydrate groups covalently attached to them. Locate and label two glycoproteins in Model 3. Explain why you would not expect any carbohydrate groups to be attached to the portion of the protein that is inside the lipid bilayer.

8. A cell membrane can vary in terms of how "fluid" or flexible it is. Describe if you would expect each of the following changes to make a particular membrane <u>more fluid</u> or <u>more rigid</u>. Explain your choices.

   a. an increase in the cholesterol content

   b. an increase in the fraction of <u>unsaturated</u> fatty-acyl tails in the membrane glycerophospholipids

9. Would you expect a reindeer living in a cold climate in northern Canada to have a cell membrane composition similar to a red-tailed deer in a warm climate in the southern United States? If so, explain why. If not, how would you expect the membranes to differ?

Model 4: Schematic diagram of different methods of transporting small molecules across the cell membrane.

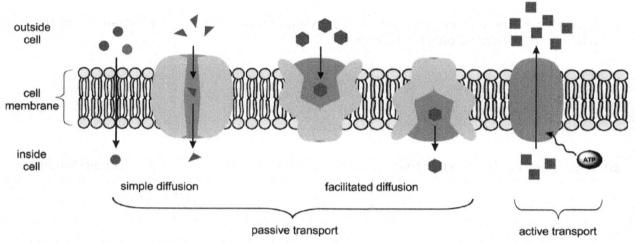

Critical Thinking Questions:

10. What are the three methods of transporting small molecules across cell membranes?

11. Are transport proteins integral or peripheral membrane proteins?

12. Based on Model 4, what does active transport require that passive transport does not?

13. Work with your team to propose one or two reasons why a cell would need to restrict what molecules can enter and/or exit.

14. Why would using a channel protein be necessary for some molecules to pass through the lipid bilayer? (Hint: consider properties of lipids.)

15. In passive transport, will molecules likely go from concentrations of **high** to **low** or **low** to **high** (circle one)?

16. Recall that osmosis is the movement of water through a membrane. Which symbol in Model 4 do you think could represent water (circles, triangles, squares, hexagons)? Explain your choice.

CA45C

17. Discuss with your team the difference between facilitated diffusion and simple diffusion based on Model 4. When you agree, describe the difference in a sentence.

18. Active transport requires the input of energy (typically ATP). Discuss with your team and propose a reason why transporting some molecules would require energy.

19. Summarize the difference(s) between active and passive transport. When would the cell use active instead of passive transport?

20. List two strengths of your team work today and why they helped your understanding.

Exercises:

1. Describe the differences between triacylglycerols and glycerophospholipids.

2. Explain why triacylglycerols would not likely be found in membranes.

3. Using the internet, find three functions of glycoproteins and describe the importance of these functions for cells.

4. Glucose is a molecule that is transported by facilitated diffusion. Propose some reasons that carefully controlling glucose transport is beneficial to cells.

5. Read the assigned pages in your text, and work the assigned problems.

Lipid Reactions
(Why do rotten foods stink?)

Model 1: Hydrolysis of a triacylglycerol

a fat molecule

Critical Thinking Questions:

1. What oxygen-containing functional group is present in a triacylglycerol? _____

2. What two functional groups are formed when a triacylglycerol is hydrolyzed by the addition of water?

3. Fill in the product(s) for the reaction in Model 1.

4. Many organisms produce enzymes that catalyze this hydrolysis reaction. Discuss with your team and propose a time or reason when it would be beneficial for an organism to break down triacylglycerols. Write your team's proposed answer below.

Model 2: Saponification of triacylglycerols

Saponification is a method for making soap that has been employed for centuries. In this reaction, sodium hydroxide (lye) and a triacylglycerol (usually animal fat or oil) are mixed together and heated. The triacylglycerol is first hydrolyzed (as in Model 1), and then the resulting fatty acids react further with NaOH.

Critical Thinking Questions

5. Fill in the product(s) in Model 2 for the first part of the reaction, the hydrolysis of a triacylglycerol with water (as in Model 1).

6. Choose one of the resulting fatty acids and write its reaction with NaOH below.

7. Label the acid and the base in the neutralization reaction in CTQ 6.

8. How is the product of this reaction different from the hydrolysis product in Model 1? Discuss with your team, and write your team answer in a sentence.

Model 3: Hydrogenation of a triacylglycerol

Hydrogenation is a process sometimes used with food products containing multiple double bonds. Partial hydrogenation will break some of the double bonds, but not all. This reaction has been used to make margarine and vegetable shortening.

Critical Thinking Questions:

9. In the reaction in Model 3 hydrogen is added to the unsaturated carbons. Does this result in **an oxidation** or **a reduction** of the triacylglycerol (circle one)?

10. Fill in the product(s) for Model 3 for a complete hydrogenation.

11. Discuss with your team how the product in Model 3 might change if the reaction was only a partial hydrogenation. Why might a company want to only partially hydrogenate oils? Propose a possible reason below.

Model 4: Oxidation of fats

Oxidation of fats results in the rancid odors of rotting food. Many foods contain antioxidants that delay this process.

Critical Thinking Questions

12. What functional group is produced from the alkene in the first oxidation step in Model 4?

13. Fill in the final products in Model 4. What functional group is produced? _____

14. Where does the oxygen come from to cause this oxidation in foods?

15. As a team, propose a method how the addition of antioxidants might help keep food fresh.

16. List the major concepts of this activity.

17. Develop a strategy for predicting the products of the reactions in this activity. Record the strategy your team developed.

18. Discuss one area for improvement in your team interactions today. Make a plan for how you can improve in this area for the next class period. Write your plan below.

Exercises:

1. A common ingredient in many prepared foods is *partially hydrogenated soybean oil.* "Partially hydrogenated" means that some (but not all) of the double bonds in the fatty acid tails have been converted to single bonds. In this process, some of the remaining *cis* double bonds are also converted to *trans* double bonds.

    a. What would be the purpose of converting some of the double bonds in the fatty acid tails to single bonds? (How would this affect the properties of the oil?)

    b. How can you reduce the amount of *trans* fats in your diet?

2. a. Complete the equation for the hydrolysis of the following triacylglycerol.

    b. Write an equation for the saponification of triacylglycerol in part (a) with NaOH.

3. Look up the structure of olestra. How does the structure differ from triacylglycerol? Why can it be used as a fat substitute?

4. Read the assigned pages in the text and work the assigned problems.

CA45D

Overview of Amino Acids and Proteins
(What does it mean for a protein to be <u>denatured</u>?)

Model 1: General structure of an α-amino acid in acidic, neutral, or basic solution.

The building blocks of proteins are α-**amino acids**, small molecules that contain a carboxylic acid and an amino group. The amino group is connected to the carbon next to the carboxyl group, designated the α carbon. There are 20 different amino acids found in proteins, differing only in the **side chain** ("R group").

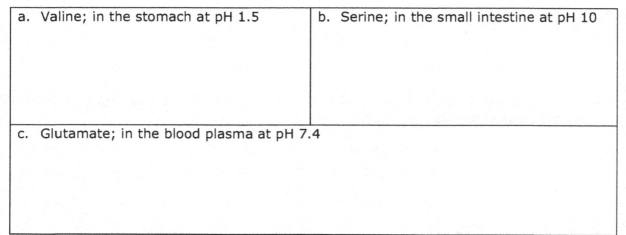

| acidic (low pH) | neutral | basic (high pH) |

Amino acids contain both a carboxylic acid (proton donor) and a basic amino group (proton acceptor). The neutral "zwitterionic" form is commonly found in aqueous solutions with a pH near neutrality. The side chain "R" may be one of 20 choices.

Critical Thinking Questions:

1. Identify the α-amino group and the carboxyl group in each structure in Model 1. Ensure all team members agree.

2. Circle the correct ionization state of an amino group

    a. at low pH: $-NH_2$ or $-NH_3^+$

    b. near neutral pH: $-NH_2$ or $-NH_3^+$

    c. at high pH: $-NH_2$ or $-NH_3^+$

3. Circle the correct ionization state of a carboxyl group

    a. at low pH: $-COOH$ or $-COO^-$

    b. near neutral pH: $-COOH$ or $-COO^-$

    c. at high pH: $-COOH$ or $-COO^-$

4. Using Models 1 and 2 (on the following page) to help you, draw the structure of the indicated amino acid as it would commonly exist under the given conditions. Charges are important.

a. Valine; in the stomach at pH 1.5	b. Serine; in the small intestine at pH 10
c. Glutamate; in the blood plasma at pH 7.4	

Model 2: Structures and abbreviations of the 20 standard amino acids as they exist at neutral pH, classified according to their side chain (R group) chemistry

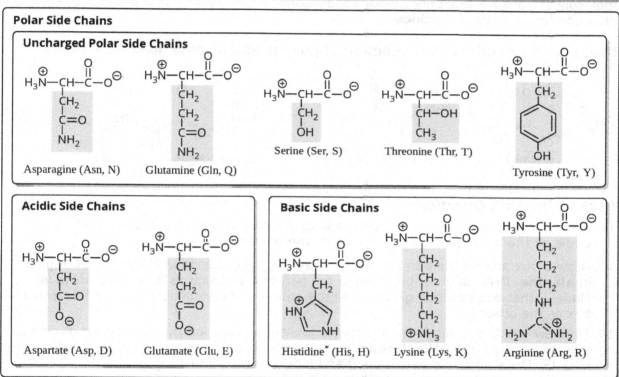

* Only a small fraction of the Histidine side chain is positively charged at neutral pH.

Critical Thinking Questions:

5. Look at the structure of alanine at the top of Model 2. Identify the amino group and the carboxyl group in alanine. Ensure that all team members agree.

6. What is the side chain of alanine (circle one)?   **methyl   ethyl   propyl   isopropyl**

7. What is the side chain of valine (circle one)?   **methyl   ethyl   propyl   isopropyl**

8. Examine the 20 amino acids, and circle each *side chain* that has N–H or O–H bonds. Which of the four classes of amino acids has the **fewest** number of side chains circled?

   Circle one: **nonpolar   uncharged polar   acidic   basic**

CA46A

9. Give the name and three-letter abbreviation of an amino acid that meets each description below.

   a. An amino acid with a polar, uncharged side chain: _____

   b. An amino acid with a polar, basic side chain: _____

   c. An amino acid containing an alcohol group: _____

   d. An amino acid with a carboxyl group in its *side chain*: _____

10. Examine the structure of tryptophan in Model 2. Discuss with your team and devise an explanation for why it is considered nonpolar even though it has an N–H bond.

11. To which of the four classes of amino acids could you apply the term *hydrophobic*?

Information:

The carboxylic acid (carboxyl) group of one amino acid and the amino group of another can undergo a condensation reaction to form a new bond, called a **peptide bond**. The new molecule is called a dipeptide. More condensations can lead to tripeptides, tetrapeptides, *etc.*, and finally to **polypeptides**.

Model 3: The condensation reaction of glycine and alanine makes a dipeptide.

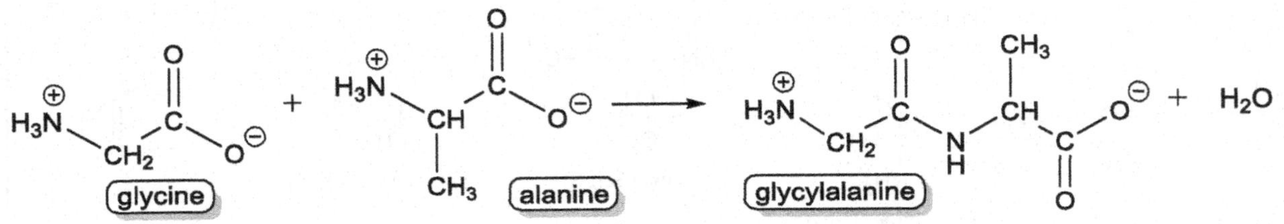

Critical Thinking Questions:

12. The α carbons are the ones bonded to both a carbonyl carbon and a nitrogen atom. Circle all the α carbons in Model 3. (Hint: there are four.)

13. Draw a box around the part of the structure for glycylalanine in Model 3 that originated in **alanine**. Draw another box around the part of the structure for glycylalanine in Model 3 that originated in **glycine**. All the atoms of glycylalanine should now be in one box or the other.

14. The new bond that was formed in the condensation reaction that connected glycine to alanine is called a **peptide bond**. This bond connects the two boxes that you drew in CTQ 12. Draw an arrow to this bond and label it in the model.

15. Consider the dipeptide Ala-Gly (alanylglycine) shown at the right. Is this the *same molecule* as Gly-Ala (glycylalanine), or are they *isomers*? If they are isomers, what type of isomers are they? Explain how you can tell. (Hint: Look at the α carbons.)

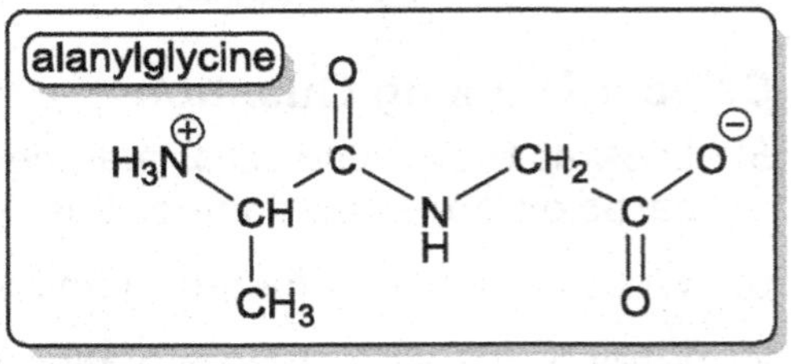

alanylglycine

16. Circle the three **side chains** in Model 4 (below). Refer to Model 2 to identify the three amino acids that condensed together to make the tripeptide, and label them in Model 4.

Model 4: A tripeptide

$$H_3\overset{\oplus}{N}-CH-\overset{O}{\overset{\|}{C}}-NH-CH-\overset{O}{\overset{\|}{C}}-NH-CH-\overset{O}{\overset{\|}{C}}-O^{\ominus}$$

with side chains: $CH-CH_3$ / CH_3 ; CH_2 / OH ; $H_2C-\overset{O}{\overset{\|}{C}}-O^{\ominus}$ / O

Model 5: Levels of protein structure

A polypeptide with a specific biological function is called a **protein**. The amino acid sequence of a protein is called the **primary structure**. Protein structure is usually classified into four levels.

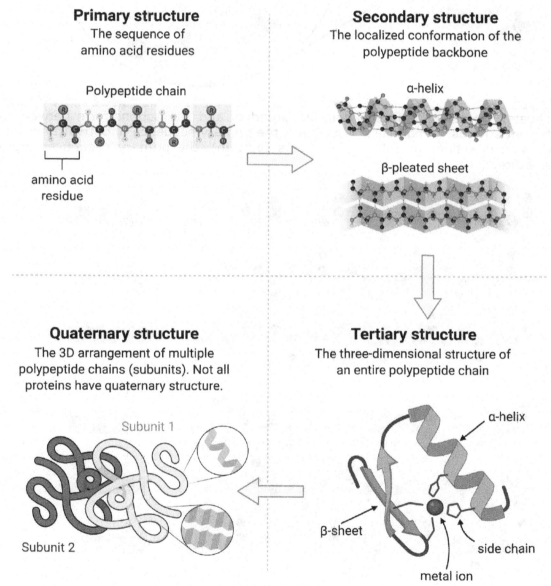

Primary structure
The sequence of amino acid residues

Polypeptide chain

amino acid residue

Secondary structure
The localized conformation of the polypeptide backbone

α-helix

β-pleated sheet

Quaternary structure
The 3D arrangement of multiple polypeptide chains (subunits). Not all proteins have quaternary structure.

Subunit 1

Subunit 2

Tertiary structure
The three-dimensional structure of an entire polypeptide chain

α-helix

β-sheet

side chain

metal ion

Created with BioRender.com

Model 6: The α helix and the β sheet are secondary structures of proteins

Secondary structures are normally held in place by *hydrogen bonding*.

In the α helix, the polypeptide backbone twists so that hydrogen bonds form four residues further along in the sequence.

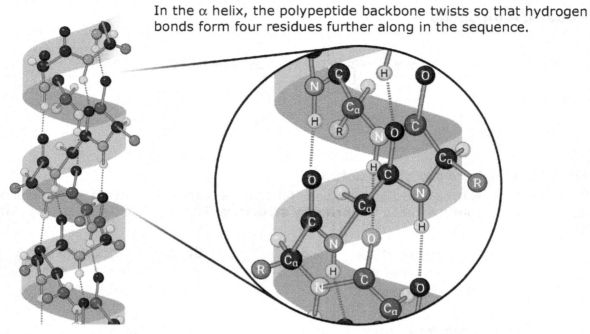

In β sheets, the polypeptide backbone is extended, and interactions form between groups on adjacent strands. In parallel sheets (top), the primary sequence of both strands proceeds in the same direction. In antiparallel sheets (bottom), the primary sequences proceed in opposite directions.

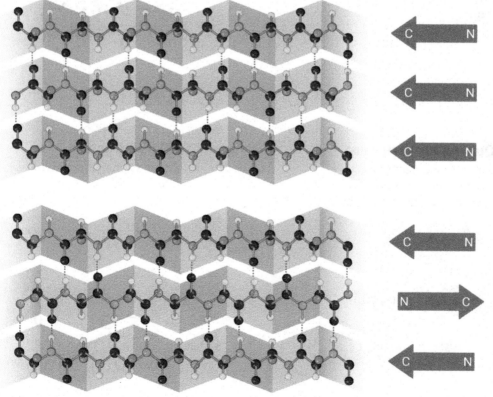

Created with BioRender.com

Critical Thinking Questions:

17. Myoglobin is a protein composed of a single polypeptide chain, while hemoglobin is composed of four polypeptides in a tetrahedral arrangement. Which one of these proteins exhibits a **quaternary** structure? _____

18. What type of interaction is indicated by the dashed lines in Model 6? _____

19. Recall that amino acids contain carboxyl groups, α-amino groups, and various side chains (R groups). Between which of these parts do the hydrogen bonds form in an α-helix? Which atoms are involved?

20. Examine the hydrogen bonding patterns in β sheets. Are the parts of the amino acids involved in hydrogen bonding in β sheets the same or different than in an α-helix? Explain in a complete sentence.

21. Label the two β sheets in Model 6 as either **parallel** or **antiparallel**.

22. What part of the protein is involved in forming secondary structure – the **backbone** containing peptide bonds or the **side chains** of the amino acids? _____

Information: Globular proteins have hydrophobic cores

The primary structure of a protein is composed of amino acids connected with covalent (peptide) bonds that are not easily broken. However, a protein must be in its *native conformation* in order to be active. The conformation is determined by the higher order structures—secondary, tertiary and quaternary. These structures are held together with weaker, noncovalent interactions.

Secondary structures (α helix, β sheet) are held in place with hydrogen bonds. Tertiary and quaternary structures also make use of hydrogen bonding, but the **most important** force driving protein folding is the **hydrophobic effect**—a process by which the water solvent attracts the polar amino acids to the outside of the protein and "squeezes" the nonpolar amino acids into the center of the structure.

Noncovalent interactions can be disturbed by heat, agitation, pH changes, detergents, salts, organic solvents, *etc.*, in a process called **denaturing**. Denatured proteins are often said to be **unfolded**, and are no longer able to perform their intended function.

Critical Thinking Questions:

23. About half of the 223 amino acid residues of the digestive enzyme trypsin are hydrophobic. Describe where in the tertiary structure you would expect to find those residues.

24. Describe where in the tertiary structure you would expect the polar amino acid residues in trypsin to be located.

25. Another digestive enzyme, chymotrypsin, has two polypeptide chains, each with 245 amino acid residues. Again, about half of the residues have nonpolar side chains. The backbone of the quaternary structure is shown below (note the two separate polypeptide chains). Describe <u>three distinct locations</u> in the overall structure you would expect the nonpolar amino acid residues to be located. Label these locations in the picture.

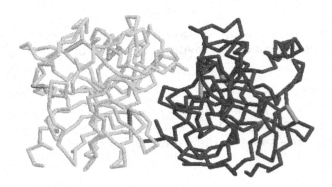

26. Work with your team to summarize the importance of hydrogen bonding in maintaining the native conformation of a protein.

27. Identify any questions remaining with your team regarding the four levels of protein structure.

Information:

Table 1: Proteins perform many important functions

Function	Type	Examples
structure	fibrous	*microfilaments* are part of the cytoskeleton
catalysis	globular	*sucrase* is the enzyme that aids the hydrolysis of sucrose to fructose and glucose
contraction	globular and fibrous	*actin* and *myosin* in muscle fibers
transport	globular	*hemoglobin* in blood; various membrane proteins perform active transport

Exercises:

1. Proteins are made of long chains of amino acids bonded together and folded into a particular shape. Proteins may be either fibrous or globular, and the specific shape of each protein is individualized to help it perform a specific function, like those in Table 1. Find three examples of proteins (in your textbook or another source) that are not listed in Table 1 and describe the type and function of the protein.

2. Explain why a peptide bond may also be called an amide bond.

3. Rank the solubility of the following amino acids in water at pH 7, from most to least soluble: Val, Ser, Phe, Lys. Explain your answer.

4. Suppose a polypeptide containing 150 amino-acid residues (chosen at random from the 20 common ones) is synthesized in the laboratory. Why is it not correct to call this polypeptide a protein?

5. A detergent, sodium dodecyl sulfate (also known as sodium lauryl sulfate, or SLS), is shown at the right. Label the hydrophilic and hydrophobic parts of the detergent.

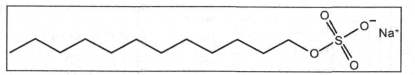

Which class of amino acid side chains would be attracted to the hydrophobic part? _____ Considering your answers to CTQs 23–25, explain how adding a detergent to a protein solution might cause the protein to denature.

6. Read the assigned pages in your text, and work the assigned problems.

CA46A

Amino Acids
(What makes each one unique?)

Model 1: Structures and abbreviations of the 20 standard amino acids as they exist at neutral pH, classified according to their side chain (R group) chemistry

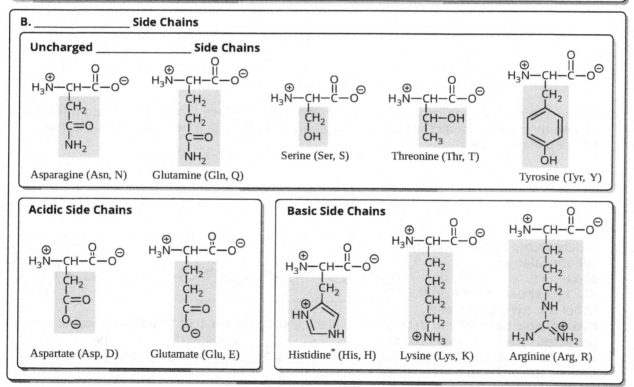

* Only a small fraction of the Histidine side chain is positively charged at neutral pH.

Critical Thinking Questions:

1. Which atom present in cysteine is different than other amino acids in group A of Model 1? _____

2. Draw the structure of valine.

3. Compare cysteine to valine. What part of the structure is similar? What part is different? Does your team agree?

4. As a team, determine the two functional groups that all amino acids have in common.

5. Why would an amino acid be called an "acid"?

Model 2: General structure of an α-amino acid

Amino acids have a similar general structure:

$$H_3\overset{\oplus}{N}-CH-\overset{\overset{\textstyle O}{\|}}{C}-O^{\ominus}$$
$$|$$
$$R$$

The carbon bonded to the carboxylic acid group is called the α (alpha) carbon, and the amino group is bonded to the α carbon. Each amino acid has unique properties due to differences in the side chains ("R groups") also attached to the α carbon. At physiological pH, the amino and carboxyl groups are charged (as indicated).

Critical Thinking Questions:

6. Consider your structure of valine in CTQ 2. **Manager:** Have each team member identify one of the following: the R group, the amino group, or the carboxylic acid group.

7. Classify the R group of valine as polar or nonpolar. Write a team consensus explanation for your choice.

8. Compare the R groups of the other amino acid structures in group A of Model 1 to valine. Are these **polar** or **nonpolar**? Fill in the blank in group A of the model with one of these terms.

9. Compare the R groups of the amino acid structures in group B of Model 1. Are these **polar** or **nonpolar**? Fill in *both* blanks in group B of the model with one of these terms.

10. Consider a water molecule. Would this molecule be **polar** or **nonpolar**? _____

11. **Circle** each **side chain** (R group) in Table 1 that can has N–H or O–H bonds and can therefore participate in hydrogen bonding with water. Which class of amino acids (**polar** or **nonpolar**) contains the most circled R groups? _____

12. Engage in a team discussion about which class or classes of amino acids you would expect to find on the exterior of proteins (interacting with the aqueous environment of the cell). Be sure to consider intermolecular forces. Work together to write a team consensus answer.

13. Which class(es) of amino acids would you expect to find on the interior of proteins? Explain.

Model 3: General structure of an α-amino acid in solutions of different pH

Critical Thinking Questions:

Refer to Model 3 to answer CTQs 14-18.

14. What is the overall **charge** of structure A? _____ Structure B? _____ Structure C? _____

15. Does the amino acid A act as an **acid** or a **base** in the forward reaction? _____

16. For the reaction from B to A (the reverse of CTQ 15), would amino acid B act as an acid or a base?

17. In a very acidic or low pH solution (when many H^+ ions are available), would an amino acid likely resemble A, B, or C? Discuss with your team and write a consensus answer. Explain your choice.

18. At a high pH, would an amino acid resemble A, B, or C? Explain your choice.

19. Using Models 1 and 3 for reference, draw the structure of the given amino acid as it would commonly exist under each of the following conditions.

  a. Valine; in the stomach at pH 1.5

  b. Serine; in the small intestine at pH 10

  c. Glutamate; in the blood plasma at pH 7.4

Information: Isoelectric points

As shown in Model 3, amino acids do not normally exist in unionized forms. When an amino acid contains both a (+) and a (−) charge but has no net charge, the structure is called a **zwitterion**.

The overall charge of an amino acid is dependent upon the pH of the solution. The amino acids with acidic or basic side chains in Model 1 contain R groups which can also change their ionization state depending on pH.

The pH at which the charge of the amino acid is neutral is called the **isoelectric point (pI)**.

Critical Thinking Questions:

20. a. Write the structure of aspartate from Model 1.

  b. Work with your team to draw the structure of aspartate at pH = 2.

  c. Write the structure of aspartate at pH = 11. Compare with your team.

21. At pH = 11, would you expect to find aspartate on the **exterior** or the **interior** of a protein? Discuss as a team and write a consensus explanation.

22. Identify at least three major concepts from this activity.

23. Summarize how the structure of an amino acid changes as the pH changes.

24. What was one strength of your team interaction today? Why was this helpful?

Exercises:
1. How does the polarity of the side chain in leucine compare to the side chain in serine?

2. Rank the solubility of the following amino acids in water at pH 7, from most to least soluble: Val, Ser, Phe, Lys. Explain your answer.

3. For the following amino acids, draw the structure of each at a pH of 2.0 and 12.0: cysteine, serine, and lysine.

4. Why do amino acids exist in their zwitterionic form at a neutral pH rather than an uncharged form?

5. Read the assigned pages in your text, and work the assigned problems.

Protein Structure
(What holds a protein's shape?)

Model 1: Peptide bond formation between glycine and alanine to make the dipeptide glycylalanine.

The acid group of one amino acid and the amino group of another can undergo a condensation reaction to form a new bond, called a **peptide bond**. The new molecule is called a dipeptide. More condensations can lead to tripeptides, tetrapeptides, *etc.*, and finally to **polypeptides**.

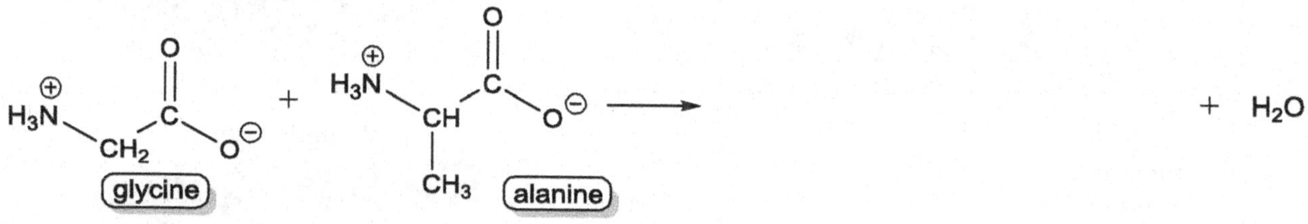

Critical Thinking Questions:

1. What small molecule is removed as part of the condensation reaction in Model 1? _____

2. Work with your team to determine the other product of the reaction in Model 1. Draw it above. What new organic functional group is formed? _____

3. Draw an arrow to the bond that was formed in the condensation reaction to connect glycine to alanine. This bond is called a **peptide bond**. Label it in the model.

4. Draw the dipeptide Ala-Gly (alanylglycine). Is this the same molecule as Gly-Ala (glycylalanine), or are they isomers? Explain.

5. Circle the three side chains (R groups) in the tripeptide shown below. Label the three amino acids (Refer to CA46B if necessary).

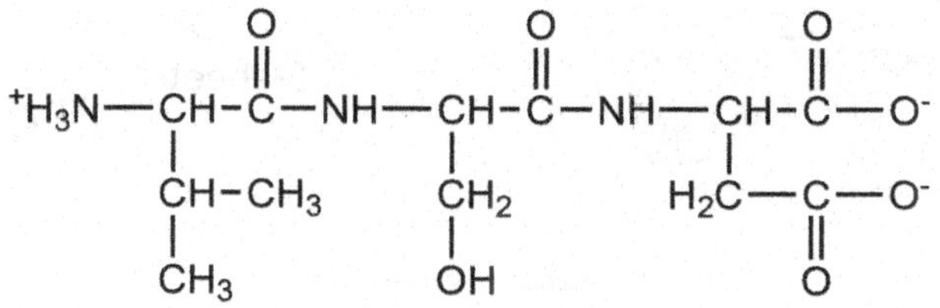

6. Discuss with your team and write a sentence that describes how the sequence of a peptide can be determined from its structure.

Model 2: Levels of protein structure

A polypeptide with a specific biological function is called a **protein**. The amino acid sequence of a protein is called the **primary structure**. Protein structure is usually classified into four levels.

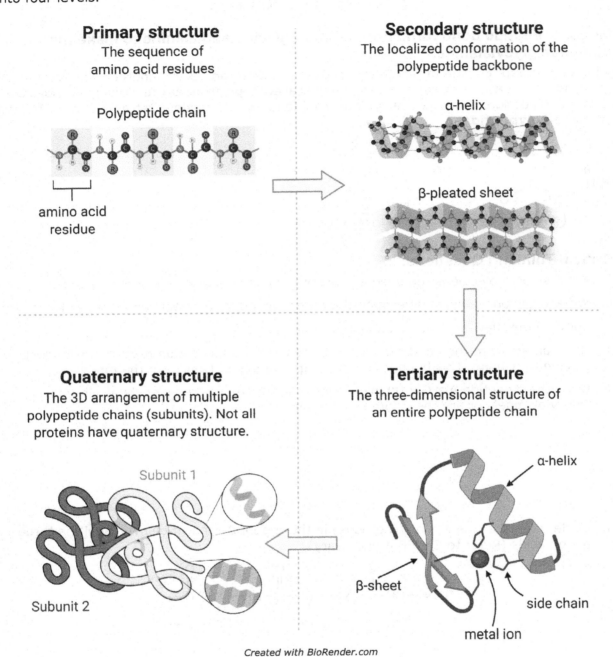

Primary structure
The sequence of amino acid residues

Polypeptide chain

amino acid residue

Secondary structure
The localized conformation of the polypeptide backbone

α-helix

β-pleated sheet

Quaternary structure
The 3D arrangement of multiple polypeptide chains (subunits). Not all proteins have quaternary structure.

Subunit 1

Subunit 2

Tertiary structure
The three-dimensional structure of an entire polypeptide chain

α-helix

β-sheet

side chain

metal ion

Created with BioRender.com

Critical Thinking Question:

7. Myoglobin is a protein composed of a single polypeptide chain, while hemoglobin is composed of four polypeptides in a tetrahedral arrangement. Which one of these proteins exhibits a **quaternary** structure? _____

Model 3: The α helix and the β sheet are secondary structures of proteins

Secondary structures are normally held in place by *hydrogen bonding*.

In the α helix, the polypeptide backbone twists so that hydrogen bonds form four residues further along in the sequence.

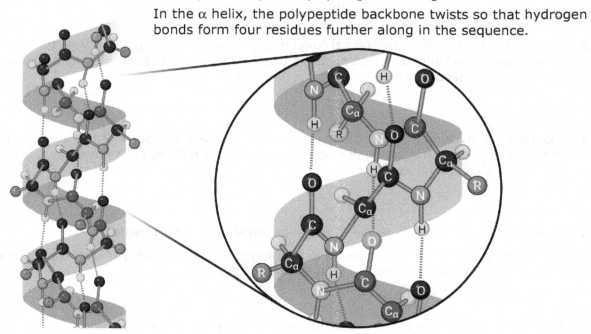

In β sheets, the polypeptide backbone is extended, and interactions form between groups on adjacent strands. In parallel sheets (top), the primary sequence of both strands proceeds in the same direction. In antiparallel sheets (bottom), the primary sequences proceed in opposite directions.

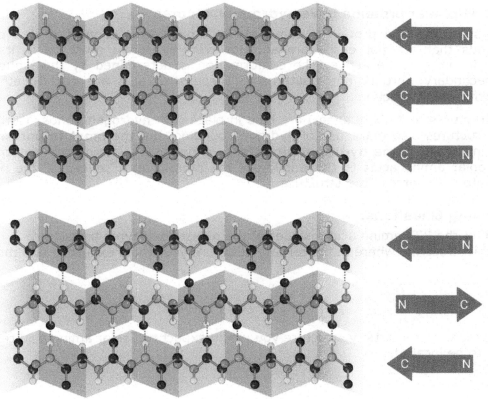

Created with BioRender.com

CA46C

Critical Thinking Questions:

8. What type of interaction is indicated by the dashed lines in Model 3?

9. Recall that amino acids contain carboxyl groups, α-amino groups, and various side chains. Between which of these parts do the hydrogen bonds form in an α-helix? Which atoms are involved?

10. Examine the hydrogen bonding patterns in β sheets. Are the parts of the amino acids involved in hydrogen bonding in β sheets the same or different than in an α-helix? Explain in a complete sentence.

11. Label the two β sheets in Model 3 as either **parallel** or **antiparallel**.

12. What part of the protein is involved in forming secondary structure – the **backbone** containing peptide bonds or the **side chains** of the amino acids?

13. How do the intermolecular forces holding secondary structure together differ from those involved in tertiary structure? Work with your team to develop a consensus explanation.

Information: Globular proteins have hydrophobic cores

The primary structure of a protein is composed of amino acids connected with covalent (peptide) bonds that are not easily broken. However, a protein must be in its *native conformation* in order to be active. The conformation is determined by the higher order structures—secondary, tertiary and quaternary. These structures are held together with weaker, noncovalent interactions.

Secondary structures (α helix, β sheet) are held in place with hydrogen bonds. Tertiary and quaternary structures also make use of hydrogen bonding, but the **most important** force driving protein folding is the **hydrophobic effect**—a process by which the water solvent attracts the polar amino acids to the outside of the protein and "squeezes" the nonpolar amino acids into the center of the structure.

Critical Thinking Questions:

14. About half of the 223 amino acid residues of the digestive enzyme trypsin are hydrophobic. Describe where in the tertiary structure you would expect to find those residues.

15. Describe where in the tertiary structure you would expect the polar amino acid residues in trypsin to be located.

16. Another digestive enzyme, chymotrypsin, has two polypeptide chains, each with 245 amino acid residues. Again, about half of the residues have nonpolar side chains. The quaternary structure is shown below (note the two separate polypeptide chains). Describe <u>three distinct locations</u> in the overall structure you would expect the nonpolar amino acid residues to be located. Label these locations in the picture, and explain why hydrophobic residues are likely to be found in those areas.

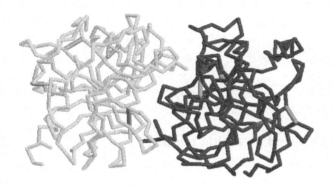

17. Work with your team to summarize the importance of hydrogen bonding in maintaining the native conformation of a protein.

18. What is one way in which your team could improve for the next class period?

Information:

Table 1: Proteins perform many important functions

Function	Type	Examples
structure	fibrous	*microfilaments* are part of the cytoskeleton
catalysis	globular	*sucrase* is the enzyme that aids the hydrolysis of sucrose to fructose and glucose
contraction	globular and fibrous	*actin* and *myosin* in muscle fibers
transport	globular	*hemoglobin* in blood; various membrane proteins perform active transport

Exercises:

1. If a mutation occurs in which a valine is replace by a serine, would you expect this to affect the protein structure? Explain.

2. Proteins are made of long chains of amino acids bonded together and folded into a particular shape. Proteins may be either fibrous or globular, and the specific shape of each protein is individualized to help it perform a specific function, like those in Table 1. Find three examples of proteins (in your textbook or another source) that are not listed in Table 1 and describe the type and function of the protein.

3. Explain why a peptide bond may also be called an amide bond.

4. a. Draw the structure of the tripeptide Ser-Lys-Asp.

   b. Would you expect to find this tripeptide on the **surface** or the **interior** of a protein?

5. Suppose a polypeptide containing 150 amino-acid residues (chosen at random from the 20 common ones) is synthesized in the laboratory. Why is it not correct to call this polypeptide a protein?

6. A detergent, sodium dodecyl sulfate (also known as sodium lauryl sulfate, or SLS), is shown at the right.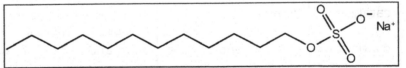
   Label the hydrophilic and hydrophobic parts of the detergent. Which class of amino acid side chains would be attracted to the hydrophobic part? _____ Considering your answers to questions 14–16, explain how adding a detergent to a protein solution might cause the protein to denature (or lose its native conformation).

7. Read the assigned pages in your text, and work the assigned problems.

Enzymes
(Why are biochemical reactions so fast?)

Model 1: Energy diagrams for exothermic reaction with and without the presence of an enzyme catalyst

a) an exothermic reaction

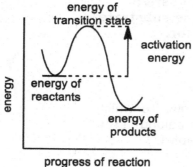

b) the same exothermic reaction in the presense of an enzyme catalyst

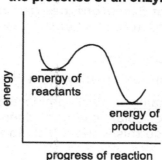

Critical Thinking Questions:

1. a. Does an enzyme change the energy of the reactants?   **yes**   or   **no**   [circle one]

    b. Does an enzyme change the energy of the products?   **yes**   or   **no**   [circle one]

2. Considering your answers to CTQ 1, does an enzyme change the *equilibrium amounts* of reactants and products? Explain.

3. Draw a vertical arrow onto Model 1 (b) that represents the magnitude of the activation energy.

4. How does the enzyme affect the rate of the reaction?

5. Work with your team to explain the function of an enzyme in a complete sentence or two.

Model 2: Enzymes bind one or more substrates into the active site

Catalysts increase the rate of a chemical reaction without being changed themselves. Most biological catalysts are protein **enzymes** that change the way a reaction takes place so that it occurs faster. The reactants in enzyme-catalyzed reactions are called **substrates**.

Enzymes lower the activation energy of a reaction by binding one or more substrates into an **active site**, using hydrophobic or hydrophilic interactions, hydrogen bonding, *etc.* This binding stretches each substrate into a reactive form and aligns it properly for the chemical reaction to take place.

Figure 1 (at right): Tertiary structure of the digestive enzyme trypsin from *Streptomyces griseus*. The substrate binds the active site, indicated with darker color.

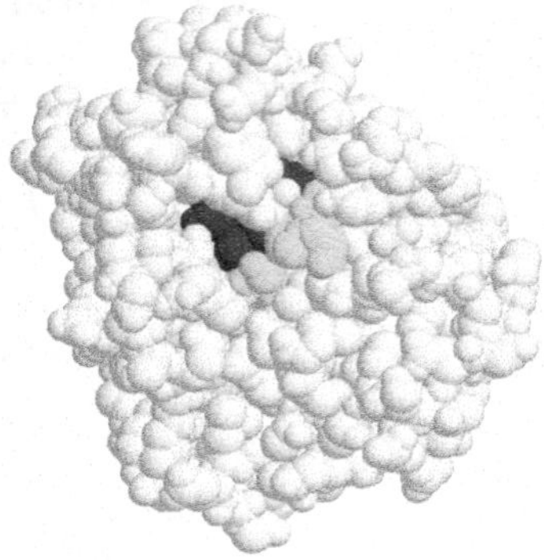

Critical Thinking Questions:

6. Circle the active site of the enzyme in Figure 1.

7. Circle the <u>side chains</u> in each amino acid below. To which class (hydrophobic, polar, charged) does each belong? (You should be able to do this without looking at a table).

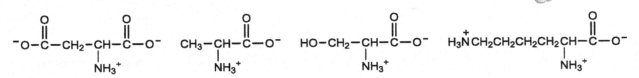

8. Suppose that aspartate (the first amino acid shown in CTQ 7) is part of a substrate. Hypothesize which of the other three amino acid side chains might be present in the active site of an enzyme in order to interact with aspartate. Work with your team to explain.

Model 3: The six classes of enzymes

Enzymes are classified according to the type of reaction they catalyze. These reactions are divided into six basic types. Enzymes are often named after their substrate and reaction class, and have the suffix "-ase." For example, the enzyme *triose phosphate isomerase* catalyzes the isomerization of glyceraldehyde 3-phosphate and dihydroxyacetone phosphate—two "triose phosphates." Since all reactions are reversible, some enzymes are named according to a product of the reaction rather than a substrate.

Find the description of the six classes of enzymes in your textbook. You are responsible for identifying reactions in the six classes. Here are some tricks to help you keep the six classes straight:

1. **oxidoreductase**: catalyzes a redox reaction; adds oxygen or removes 2 hydrogen atoms from substrate; requires a cofactor such as NAD^+, $NADP^+$ or FAD;

2. **transferase**: transfers a functional group (such as NH_2 or phosphate) between substrates;

3. **hydrolase**: catalyzes a hydrolysis reaction; substrate + $H_2O \rightarrow$ two products;

4. **lyase**: adds or removes groups involving a double bond (no ATP required);

5. **isomerase**: makes an isomer of the substrate by rearrangement;

6. **ligase**: forms a bond to join two substrates using ATP hydrolysis for energy.

Critical Thinking Questions:

9. If an oxidoreductase catalyzes the removal of two hydrogen atoms from a substrate, is the substrate oxidized or reduced?

10. Work with your team to answer the questions about the following reaction.

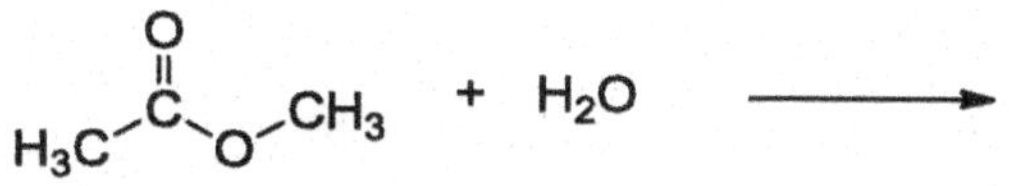

a. Identify the organic functional group present in the substrate. _____

b. What is the type of organic reaction in which water is added to split an organic molecule into two parts?

c. Complete the reaction. List the functional groups generated in the products of the reaction.

d. Which of the six classes of enzyme would catalyze this reaction? _____

11. Which category of enzyme would catalyze an addition of water across a carbon-carbon double bond as shown?

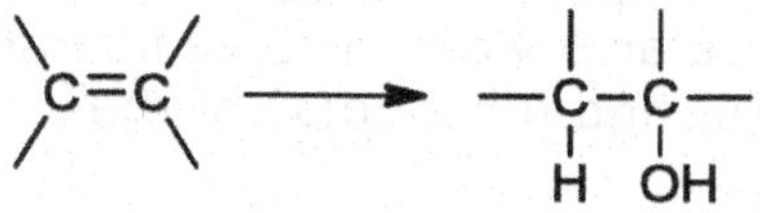

12. Which of the six classes of enzymes catalyze each of the following reactions?

a.

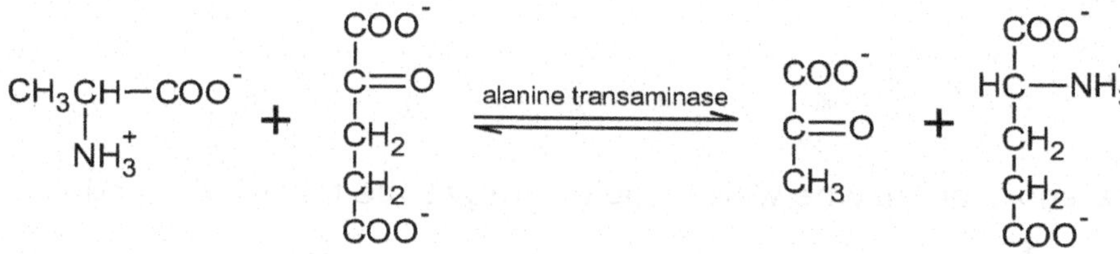

b.

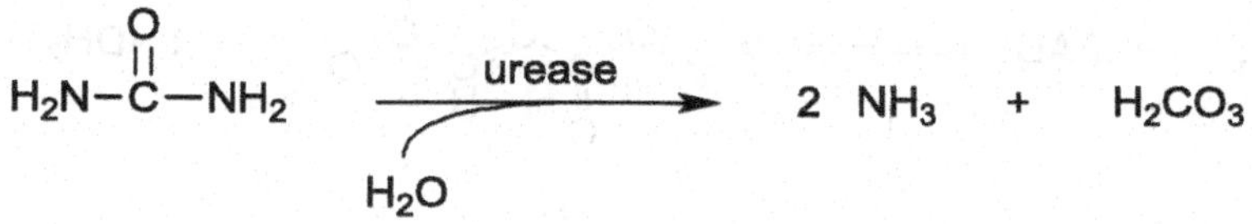

c.

d.

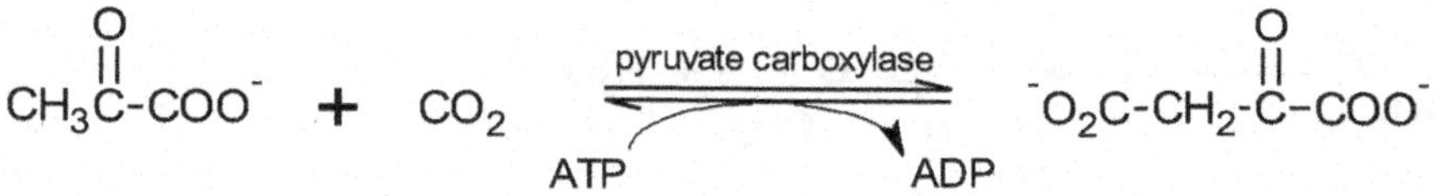

e.

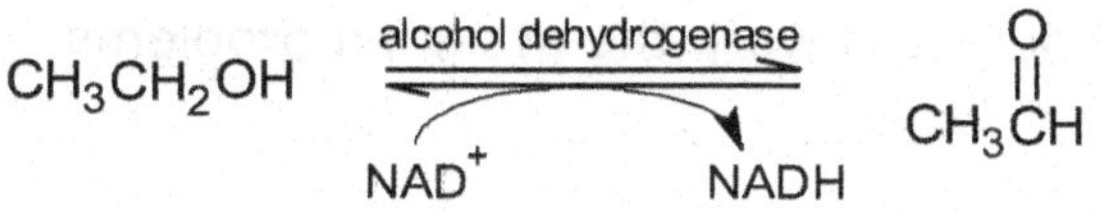

f.

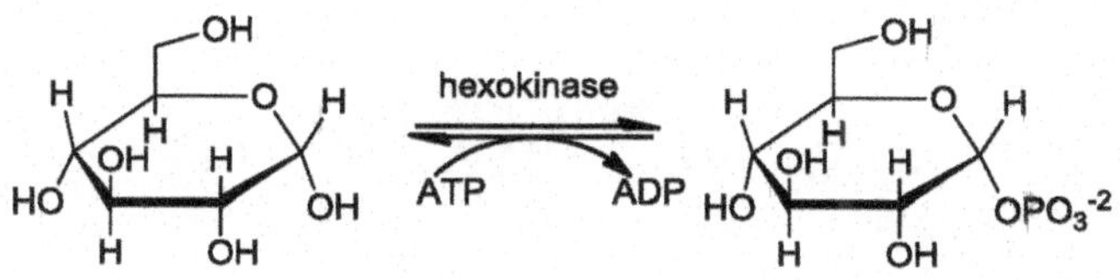

13. What are the names of the two substrate molecules shown for hexokinase in CTQ 12f? Be specific.

14. What are the major concepts of this activity?

15. List any questions remaining with your team about enzymes or their classification.

Exercises:

1. Often, an enzyme requires a cofactor or prosthetic group in its active site in order to be an efficient catalyst. Many such cofactors are synthesized by plants and obtained in the diet as vitamins. Find a description in your text of a particular B vitamin. Give its name and describe its biological function in terms of the name(s) and function(s) of the enzyme(s) that require it.

2. Identify the category of enzyme which would catalyze the following reactions.

a. + FAD $\rightleftharpoons$ + FADH$_2$

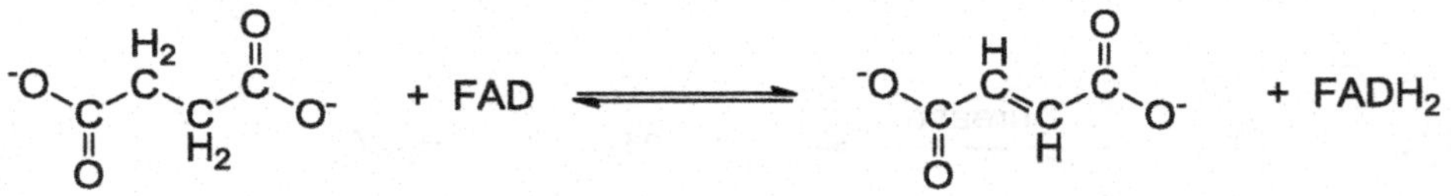

b. $^+H_3N-\overset{H}{\underset{|}{C}}-COO^- + H_2O \longrightarrow {}^+H_3N-\overset{H}{\underset{|}{C}}-COO^- + HPO_4^{2-}$
with CH_2 / OPO_3^{2-} on left and CH_2 / OH on right

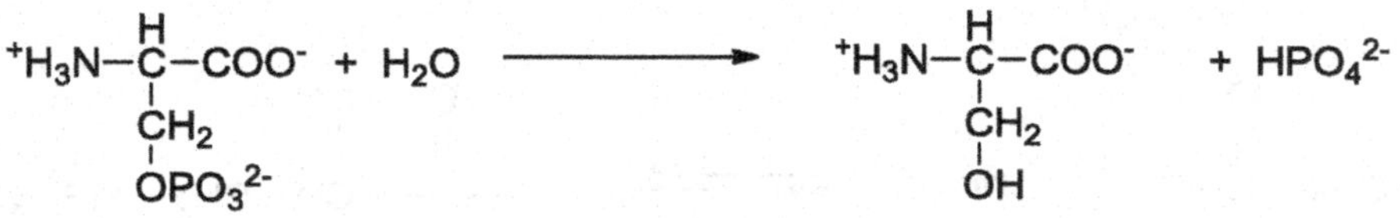

c. $CH_3\overset{OH}{\underset{|}{C}}HCH_3 \longrightarrow \overset{H}{\underset{H}{\diagdown}}C=C\overset{H}{\underset{CH_3}{\diagup}} + H_2O$

3. Read the assigned pages in your textbook, and work the assigned problems.

Effects on Enzyme Activity
(How does enzyme activity change?)

Model 1: Effects of changes in concentrations on reaction rate

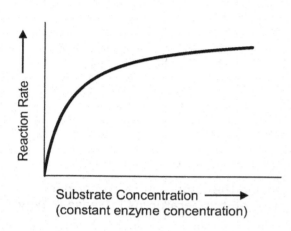

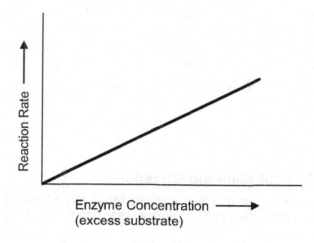

(a) Change of reaction rate with increasing substrate concentration and constant enzyme concentration.

(b) Change of reaction rate with increasing enzyme concentration and excess substrate.

Critical Thinking Questions:

Manager: Rotate team members to offer the first explanation for CTQs 1–3.

1. In Model 1(a), which concentration changes: the **enzyme** or **substrate** (circle one)?

2. What happens to the reaction rate in (a) when more substrate is added?

3. Why does the rate in (a) continue to go up as more substrate is added?

4. Work with your team to propose a consensus reason why the rate in (a) levels off.

5. In Model 1(b), which concentration changes: the **enzyme** or **substrate** (circle one)?

6. *Manager:* Lead a team discussion about why the rate in (b) continues to increase and does not level off like the one in (a). Write a consensus explanation.

Model 2: Effect of temperature on enzyme reaction rates

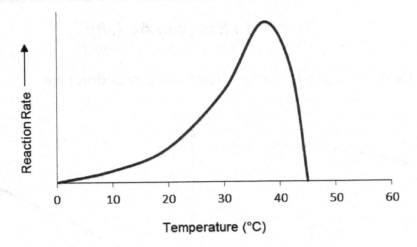

Critical Thinking Questions:

7. Work with your team to recall the effect of increasing the temperature on normal (uncatalyzed) chemical reactions. What happens when temperature is increased? Why?

8. Based on Model 2, what about enzyme-catalyzed reactions is similar to uncatalyzed chemical reactions as the temperature increases? What is different?

9. Discuss as a team to propose a reason that reaction rates for enzyme-catalyzed reactions rapidly fall off at higher temperatures. *Manager:* Ensure that everyone participates in the discussion. Write your proposal below.

Model 3: Effect of pH on enzyme reaction rates

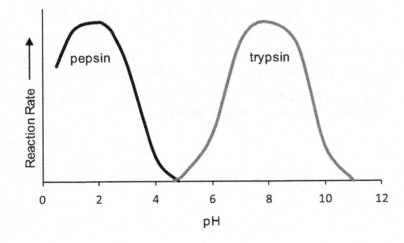

Critical Thinking Questions:

10. What is the optimal pH for pepsin activity? _____ Trypsin activity? _____

Manager: Lead the discussion for CTQ 11.

11. a. How does a change in pH affect individual amino acids?

    b. When an amino acid structure is part of a polypeptide chain, what part of the structure can be affected by pH changes?

    c. Which class(es) of amino acids would be most affected by a change in pH?

12. Discuss as a team and describe what you think happens to the three dimensional shape of pepsin at a pH of 4.

13. Why is there an optimal pH for enzyme activity?

14. Summarize, in general, what happens to an enzyme when it encounters conditions that are not optimal for its activity. **Manager:** Ensure everyone understands.

15. Describe one aspect of enzyme denaturation that remains unclear.

16. Give an example how the manager's role today helped ensure everyone understood the material.

17. What was one strength of your team today and why was it beneficial for completion of the activity?

Exercises:

1. Review Model 3. How will the rates of a pepsin-catalyzed reaction at a pH of 1 be different than at a pH of 4?

2. Chymotrypsin, a protease important for digestion in the small intestine, has an optimal pH of 7.8. What would happen to the activity of this enzyme if were moved to the acidic environment in the stomach?

3. *Thermus aquaticus* is a bacteria found in hot springs. A student attempts to purify an enzyme from this bacteria; however, after her purification, her tests showed no enzyme activity. Wondering if her results indicate a mistake in her purification process, she asks you for advice. What questions would you ask her? What advice would you give?

4. Read the assigned pages in your textbook and work the assigned problems.

Nucleic Acids
(What is DNA made of?)

Model 1: Structure of nucleotides

Nucleic acids are polymers made from smaller building blocks. The two main types of nucleic acids are **r**ibo**n**ucleic **a**cid (RNA) and 2'-**d**eoxyribo**n**ucleic **a**cid (DNA).

The building blocks of RNA and DNA are nucleotides (see Figure 1). By convention, the carbons in the sugar residue are numbered with "primes"—1', 2', 3', 4', and 5' (read as "one prime," "two prime," *etc.*) to distinguish them from the numbers of the carbons in the base.

Figure 1: Nucleotides consist of a nitrogen-containing base bonded to a sugar residue and at least one phosphate group.

Figure 2: The organic molecules (bases) pyrimidine and purine

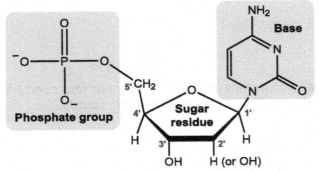

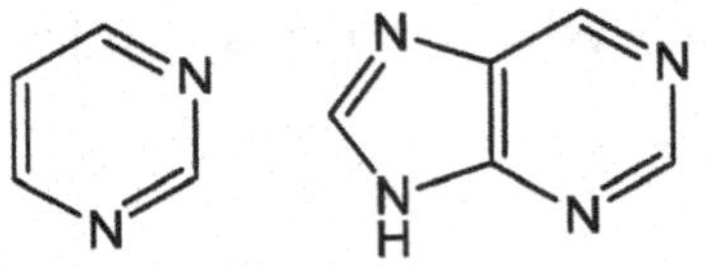

pyrimidine   purine

Figure 3: The nitrogen-containing bases in DNA and RNA

Adenine (A)
DNA and RNA

Guanine (G)
DNA and RNA

Cytosine (C)
DNA and RNA

Thymine (T)
DNA only

Uracil (U)
RNA only

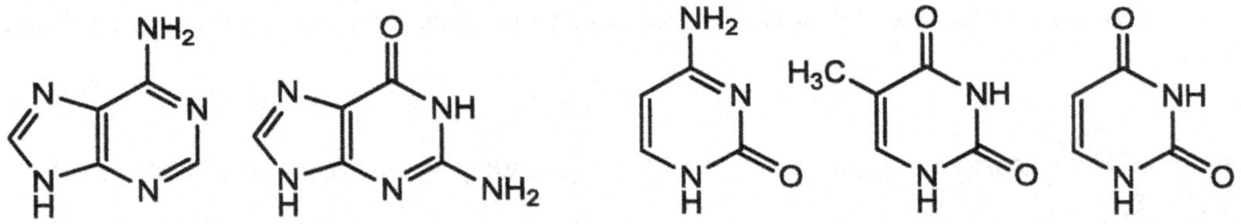

Critical Thinking Questions:

1. What are the three components of a nucleotide?

2. Which bases are larger in size: **purines** or **pyrimidines** [circle one]?

3. Based on Figures 2 and 3, which nitrogen-containing bases in DNA and RNA are purines? Which are pyrimidines?

4. The group bonded to the 2' position of the sugar residue (−H or −OH) determines whether the sugar is *ribose* or *deoxyribose*. Discuss with your team which of these groups would be present in *deoxyribose*. Explain your choice.

5. Which base(s) are unique to DNA? _____

6. Which base(s) are unique to RNA? _____

7. Which **base** is shown as part of the nucleotide in Figure 1? _____

Model 2: The primary structure of nucleic acids is the base sequence.

Similar to the levels of protein structure that we have seen, nucleic acids also have primary, secondary, and tertiary structures. The **primary structure** is the nucleic acid sequence. The nucleotides are connected from the 5' phosphate on one nucleotide to the 3' hydroxyl group on the next, *via* a **phosphodiester** linkage (see Figure 4).

Figure 4: Nucleotides condense together *via* phosphodiester linkages.

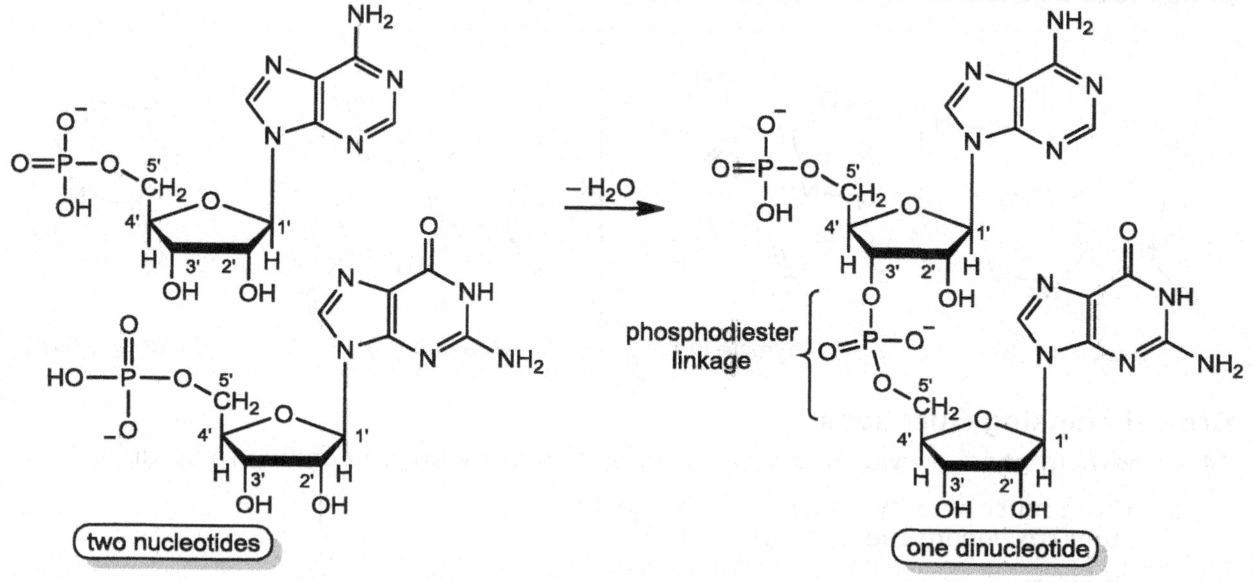

Critical Thinking Questions:

8. In Figure 4, **circle** an −OH group and an H in the reactant molecules that are lost as a molecule of water in the condensation reaction.

9. Are the nucleotides in Figure 4 **RNA** or **DNA** (circle one)? Explain how you can tell.

10. The **5' end** of a polynucleotide has a free 5'-phosphate (meaning the phosphate is not bridging two nucleotides). In the dinucleotide product in Figure 4, is the 5' end shown at **the top** or **the bottom** (circle one)? Name the base at the 5' end.

11. The **3' end** of a polynucleotide has a free 3'-hydroxyl group, meaning there is no phosphate group attached to it. In the dinucleotide product in Figure 4, is the 3' end at **the top** or **the bottom** (circle one)? Name the base at the 3' end.

12. Like polypeptide chains, nucleotide chains have directionality. By convention, base sequences are written using the one letter codes in the 5' → 3' direction. Using this directionality, what is the base sequence of the dinucleotide in Figure 4?

13. When the base adenine is attached to the sugar ribose, it is called *adenosine*, and if a phosphate group is attached, it is adenosine monophosphate (abbreviated AMP). Circle the correct name for the nucleotide at the **top** of the reactants in Figure 4. Ensure that all team members understand why this is the correct choice.

   o   adenine-5'-monophosphate
   o   adenosine-5'-monophosphate

   o   adenine-3'-monophosphate
   o   adenosine-3'-monophosphate

Model 3: Base pairing of nucleotides creates secondary structures.

The most important **secondary structure** of DNA is the famous double helix. In this structure, two sugar-phosphate backbones spiral around each other, and are held together by **hydrogen bonds** between pairs of nitrogen-containing bases, or **base pairs**. Figure 5 illustrates how the hydrogen bonding partnership works. Note that bases almost never pair with the wrong partner, because the hydrogen bonds would not line up correctly.

Figure 5: Hydrogen bonds hold AT and CG (complementary) base pairs together in DNA. The shaded "R" indicates where the bases are attached to the sugar-phosphate backbone. In RNA, uracil replaces thymine.

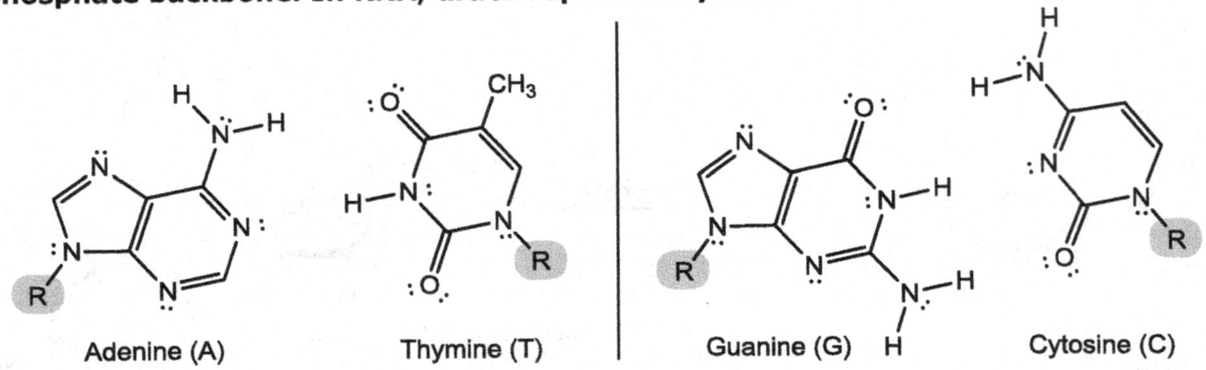

Adenine (A)   Thymine (T)   Guanine (G)   H   Cytosine (C)

Critical Thinking Questions:

14. Review: In general, two conditions are required for hydrogen bonding to occur:

    a. There must be a hydrogen atom bonded to a _____ or _____ atom (to *donate* the hydrogen bond).

    b. There must be a _____ (to *accept* the hydrogen bond).

15. Use dashed lines to draw hydrogen bonds between A and T in the appropriate locations in Figure 5. Do the same for the CG pair.

16. DNA double helices with more CG content are more heat stable than those with more AT content. Explain why this is so. (See Figure 5.)

17. Words or phrases that read the same forwards and backwards are called palindromes. Examples are "radar" or the phrase "A man, a plan, a canal: Panama!" Suppose that one strand of DNA is palindromic (such as ATGGTA). Would the complementary strand also be palindromic? (Hint: Write the complementary sequence.) Explain.

Model 4 (optional): Tertiary structure of nucleic acids

Individual strands of RNA or DNA can also have tertiary structure. Consider the typical RNA molecules shown in Figure 6.

Figure 6: Secondary (a) and tertiary (b) structures of typical transfer RNAs.

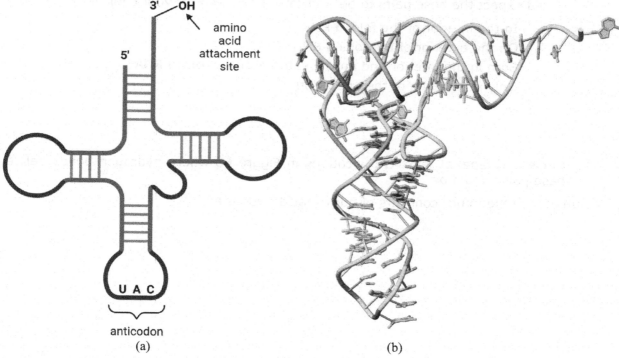

(a)   (b)

[Figure 6(a) created with Biorender.com. Figure 6(b): Image of 1EHZ (Shi H, Moore PB. The crystal structure of yeast phenylalanine tRNA at 1.93 A resolution: a classic structure revisited. RNA. 2000 Aug;6(8):1091-105) created withJmol: an open-source Java viewer for chemical structures in 3D. http://www.jmol.org/

Critical Thinking Questions:

18. a. The three-base anticodon makes hydrogen bonds to messenger RNA. Circle the **anticodon** part of the transfer RNA (tRNA) in Figure 6a.

   b. Circle the **free 3' end** of the tRNA in Figure 6a. Explain its function.

   c. During the life of a cell, the information in DNA is copied to messenger RNA, which contains the information needed to synthesize each protein that the cell requires. Each transfer RNA specifically attaches to a particular amino acid (in Figure 6a, it would be valine). What would be the minimum number of different tRNAs required to make proteins with all the different amino acids? Explain.

19. Consider the tRNA in Figure 6b, in which the hydrogen atoms are *not* shown:

   a. Describe how the ribose and phosphate backbone is depicted in Figure 6b.

   b. Choose the location from the list that indicates where in the tertiary structure you would expect the base pairs to be. Indicate the reason for your choice.

   o   In the interior of the folded RNA

   o   On the exterior of the folded RNA

   o   Both in the interior and on the exterior of the folded RNA

   o   Only in the area of the anticodon.

   c. Circle and label at least two locations in Figure 6b where hydrogen bonds between base pairs would be.

20. Summarize the major concepts of nucleic acid structure.

21. Explain how your team ensured that each member understood the material as you worked through the activity today.

Exercises:

1. Identify each species as either a nitrogenous base or a nucleotide.

   a. adenine *nitrogenous base, RNA and DNA*

   b. guanosine-5'-triphosphate *nucleotide, RNA and DNA*

   c. T *nitrogenous base, DNA*

   d. UDP *nucleotide, RNA*

2. For each species in Exercise 1, tell whether it would be found in RNA, DNA, or either.

3. Identify each base as a purine or a pyrimidine.

   a. cytosine *Pyrimidine*

   b. hypoxanthine, *Purine*

4. Read the assigned pages in your textbook, and work the assigned problems.

DNA Replication
(How can DNA copy its information?)

Model 1: Double stranded DNA in the ladder formation (a) and a double helix (b)

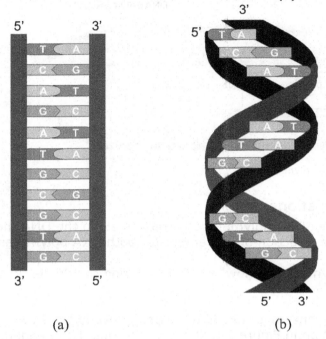

(a)   (b)

Critical Thinking Questions:

1. What is represented by the dark ribbon-like structures on the outside of Figure 1a?

2. What is represented by the letters (the rungs of the ladder)?

3. How are the rungs of the ladder held together?

Information:

By convention, DNA is written as a *sequence* of bases from 5' → 3'. This is known as the DNA **informational strand**. The complementary 3' → 5' strand (the DNA **template strand**) can be determined if needed based on base pairing.

4. a. Write the informational strand sequence from the DNA in Model 1 (a). Compare your answer with your team.

   5'   3'

   b. Write the complementary template strand sequence.

   3'   5'

5. Write the template strand for the informational strand sequence below.

   5' A G C A A T C G T T A G C G 3'

CA48B

Model 2: Replication

DNA replication produces two daughter strands—copies of the original double-stranded DNA. Each initial DNA strand (parent strand) serves as a template for the formation of a new strand based on complementary base pairing.

Figure 2: Schematic diagram of DNA replication in *E. coli*

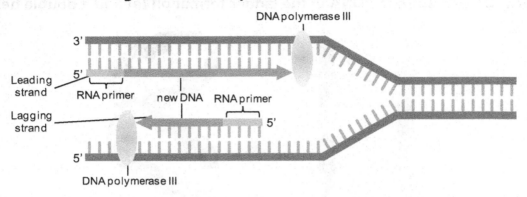

Critical Thinking Questions:

6. The enzyme complex DNA polymerase III moves along the DNA, helping to create a new strand. Indicate the direction of movement for both DNA polymerases shown in Figure 2.

7. How does DNA polymerase "know" which nucleotide to add next?

8. Replication requires enzymes called helicases to unwind the DNA. Work with your team to identify the location in Figure 2 that a helicase would be needed. Draw a square on the diagram to represent the helicase.

9. What holds the new nucleotides (on the leading and lagging strands) in place?

10. Recall what the dark ribbon structure represents in Figure 2 (CTQ 1). What needs to happen after the nucleotides are added for the new DNA to create this structure? (Review CA 48A)

11. a. Which strand is not continuous in its replication?

    b. DNA polymerase III has several important roles for replication. The polymerase can only move in a 5'→3' direction. Work with your team to determine why replication does not proceed continuously on both strands.

12. DNA polymerase requires that a short RNA primer (4–10 bases) bind with the DNA in order to begin replication. In the example diagram below, two primers are available: 5'-UACA-3' and 5'-AACG-3'.

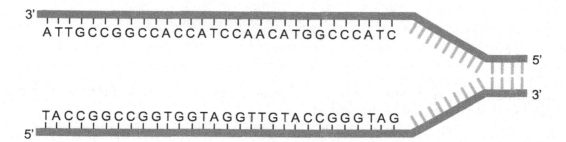

3' A T T G C C G G C C A C C A T C C A A C A T G G C C C A T C   5'

                                                                        3'

    T A C C G G C C G G T G G T A G G T T G T A C C G G G T A G
5'

   a. Work with your team to add primers into the appropriate locations in the DNA above (U can base pair with A).

   b. Complete the sequence of DNA replication performed by DNA polymerase on each strand.

   c. Draw a circle onto the diagram to represent DNA polymerase III.

   d. Indicate the location on the lagging strand which would not be replicated in this fragment. Describe what will need to happen in order to replicate this section.

   e. In your diagram, both daughter strands now have four bases of RNA at the 5' end. What would have to happen in order to form two complete daughter strands?

   f. Ligase is an enzyme that will repair any nicks or breaks left in the DNA backbone. Describe where this enzyme will be needed to create a continuous chain, and label this place in the diagram. Ensure that all team members agree.

13. Discuss with your team which you think would be the most detrimental to replication: a mutation in helicase, DNA polymerase, or ligase? Choose one, and explain your choice.

14. List the major concept(s) of this activity.

15. Write any questions remaining with your team about DNA structure and/or replication.

Exercises:

1. What is meant by complementary base pairing? Why is it important in the replication process?

2. In human DNA, replication begins at multiple positions at the same time. Propose a reason that this would be beneficial.

3. Use the internet to research the differences between DNA polymerase I and III in *E. coli*. What role does each play in replication? Why is the role important?

4. Research to find the answer to the following questions: What are Okazaki fragments? How are they joined together? How did they get their name?

5. Read the assigned pages in your textbook, and work the assigned problems.

Transcription and Translation
(How are proteins made?)

Information:

- Messenger RNA (mRNA) – carries instructions from DNA for protein synthesis
- Ribosomal RNA (rRNA) – combines with proteins to form **ribosomes** – the site of protein synthesis
- Transfer RNA (tRNA) – delivers amino acids to ribosomes

Critical Thinking Questions:

1. Discuss with your team to recall the two ways RNA differs structurally from DNA, and record the team consensus.

2. a. mRNA carries information from DNA for protein formation. In which part (organelle) of the cell would mRNA be formed?

   b. In which part of the cell would mRNA move to?

3. In which part of the cell would tRNAs be found?

4. In order to form mRNA from DNA, the DNA double helix must first be uncoiled. Because RNA is single stranded, only one strand of the DNA is **transcribed** (copied) into mRNA. This process is directed by RNA polymerase. Complete the following sequences. Compare your answers with your team. (Recall the differences from CTQ 1).

   DNA informational strand: 5′ ATG   CCA   GTA   GGC   CAC   TTG   TCA   3′

   DNA template strand: 3′   5′

   mRNA: 5′   3′

5. Which of the two DNA strands' sequences in CTQ 4 dictates the sequence of mRNA?

Model 1: The Genetic Code

An mRNA sequence is **translated** into the primary structure of a protein. A sequence of three ribonucleotides (a **codon**) codes for one amino acid of a protein.

Table 1. The 64 codons in the genetic code and their associated amino acids

First Position (5' end)	Second Position				Third position (3' end)
	U	C	A	G	
U	Phe	Ser	Tyr	Cys	U
	Phe	Ser	Tyr	Cys	C
	Leu	Ser	Stop	Stop	A
	Leu	Ser	Stop	Trp	G
C	Leu	Pro	His	Arg	U
	Leu	Pro	His	Arg	C
	Leu	Pro	Gln	Arg	A
	Leu	Pro	Gln	Arg	G
A	Ile	Thr	Asn	Ser	U
	Ile	Thr	Asn	Ser	C
	Ile	Thr	Lys	Arg	A
	Met	Thr	Lys	Arg	G
G	Val	Ala	Asp	Gly	U
	Val	Ala	Asp	Gly	C
	Val	Ala	Glu	Gly	A
	Val	Ala	Glu	Gly	G

Critical Thinking Questions:

6. If the mRNA codon is CCG, what is the amino acid?

7. What are the three stop codons?

8. Which amino acid(s) has(have) the most codons?

9. Work with your team to propose a reason why having multiple codons for some amino acids would be beneficial.

10. AUG is the only start codon. What amino acid does it code for?

11. Complete the following for a DNA fragment. Compare your answers with your team.

    DNA informational strand: 5' ATG GCT CAA TAT GTT GTC CGA   3'

    DNA template strand: 3'   5'

    mRNA: 5'   3'

    Protein sequence: N-term   C-term

Model 2: Schematic representation of a typical tRNA Structure.

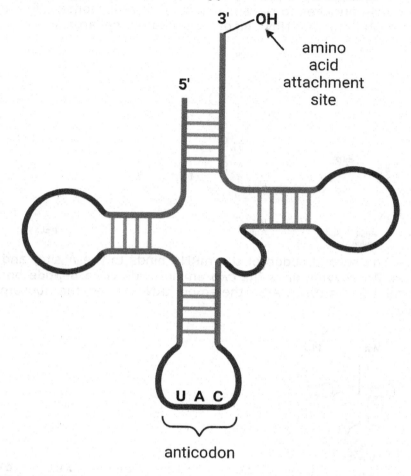

anticodon

Critical Thinking Questions:

12. What weak interactions are responsible for maintaining tRNA structure?

13. a. With what will the anticodon region of the tRNA in Model 2 interact?

    b. The mRNA codon would bind to (under) the tRNA anticodon in Model 2 in a 5′ → 3′ orientation. What would the codon sequence be?

    c. What amino acid would be attached to the top of the tRNA?

14. a. If an mRNA codon is CGA, what is the corresponding anticodon on the tRNA?

    b. Which amino acid would be attached to the tRNA?

Model 4: Translation from mRNA to protein

- ***Initiation*** – mRNA attaches to the small subunit of a ribosome, allowing the first codon to bind to the P site. The first tRNA binds, followed by the large subunit.

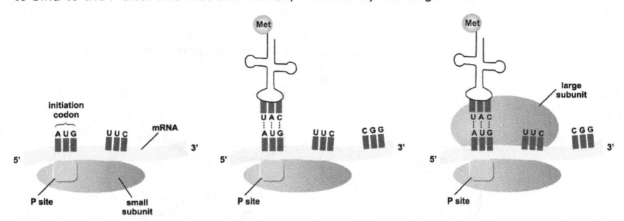

- ***Elongation*** – The second codon of the mRNA binds to the A site, and the appropriate tRNA interacts. An enzyme links the two amino acids via a peptide bond, releasing the first amino acid from its tRNA. After the empty tRNA leaves, the ribosome shifts, and the process repeats.

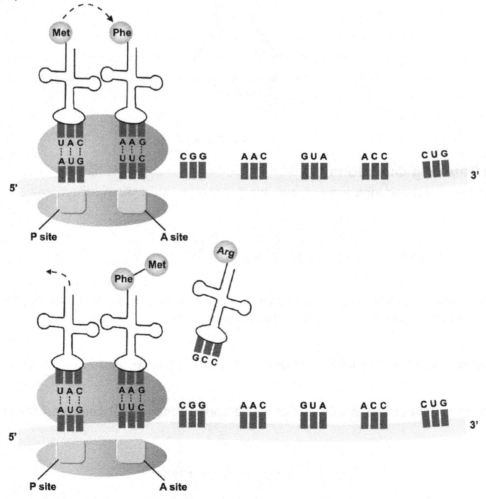

- ***Termination*** – The polypeptide continues to grow until a STOP codon is reached. An enzyme then cleaves the polypeptide from the last tRNA. The two ribosome subunits separate.

Critical Thinking Questions:

15. a.   What is the first codon of mRNA in Model 4 that binds to the ribosome?

    b.   What is the anticodon of the tRNA?

    c.   What amino acid is the first one of the protein?

16. Review Model 4. What happens to the ribosome once the first tRNA attaches?

17. a.   What is the second codon of the mRNA in Model 4?

    b.   To which site on the ribosome does it bind?

18. Discuss with your team how the two amino acids attached to individual tRNAs link together to begin forming the polypeptide. Write your consensus explanation.

19. What happens to a tRNA once its amino acid is removed?

20. As a team, discuss how a new tRNA accesses the A site. Indicate what must happen for the A site to become available.

21. What is the polypeptide sequence represented by the complete mRNA sequence shown in Model 4? Compare your answer with your team.

22. How does translation stop?

23. Work with your team to write a one-sentence summary of the role that tRNA plays in translation.

24. What is(are) the major concept(s) of this activity?

25. What is one strength of your team work today? Why is this a strength?

Exercises:

1. Describe two specific roles that enzymes play in translation.

2. What peptide is sequenced from the following DNA informational strand? First write the template and the mRNA sequences.

   5' – T-C-A-G-G-C-T-A-C-A-C-A – 3'

3. What mRNA sequence would give rise to the protein segment Phe-His-Pro-Val? What is the informational strand sequence? How many different DNA sequences would give this same peptide?

4. For the mRNA sequence in Exercise 3, draw a picture showing the corresponding tRNA interacting with the mRNA for the amino acid Phe. Include the anticodon and the appropriate amino acid.

5. Read the assigned pages in your textbook, and work the assigned problems.

Overview of Metabolism
(How do we get energy from food?)

Model 1: Breakdown of glucose

$$C_6H_{12}O_6 + O_2 \longrightarrow CO_2 + H_2O + \text{energy (673 kcal/mol)} \qquad (1)$$

Critical Thinking Questions:

1. Add coefficients to balance reaction (1) in Model 1.

2. Does reaction (1) **release** or **absorb** energy? _____

   Is the reaction **exothermic** or **endothermic**? _____

3. Reaction (1) is a redox reaction. Work with your team to identify the atom that is reduced and the atom that is oxidized in this reaction. (Recall that when in a molecular compound, oxygen has an oxidation number of -2 and hydrogen has a +1).

   Oxidized: _____   Reduced: _____

Model 2: Breakdown of a fatty acid

The less oxidized a molecule is, the more energy it has available to be released when it undergoes oxidation.

$$C_{16}H_{32}O_2 + O_2 \longrightarrow \qquad\qquad\qquad\qquad (2)$$

Critical Thinking Questions:

4. Like carbohydrates, fatty acids are metabolized to carbon dioxide and water. Complete and balance reaction (2) in Model 2.

5. Apply the shortcut method used with organic molecules to determine which atom is oxidized and which is reduced in reaction (2).

   Oxidized: _____   Reduced: _____

6. The glucose in reaction (1) has 6 carbons and 6 oxygens, while the fatty acid in reaction (2) has 16 carbons and only 2 oxygens. Which molecule is in a less oxidized (more reduced) state? Discuss with your team and write a consensus explanation.

7. Would you expect reaction (2) to release more or less energy than reaction (1)? Explain.

8. The oxidation of the fatty acid in Model 2 actually releases 2256 kcal/mol. Does this agree with your prediction in CTQ 7?

9. How much energy would be required to **synthesize** the same molecule in the reverse reaction?

Model 3: Overview of Metabolism

anabolism – synthetic (reductive) metabolic reactions which require energy
catabolism – degradative (oxidative) metabolic reactions which produce energy

Metabolic processes are all those related to the production and use of energy. In catabolism, complex carbohydrates, proteins, and lipids must be broken down into their component molecules (as shown in the diagram) in order to be absorbed into the bloodstream. These molecules make their way into the cells *via* the lymphatic and circulatory systems. Once inside cells, they can be broken down (oxidized) and their energy stored.

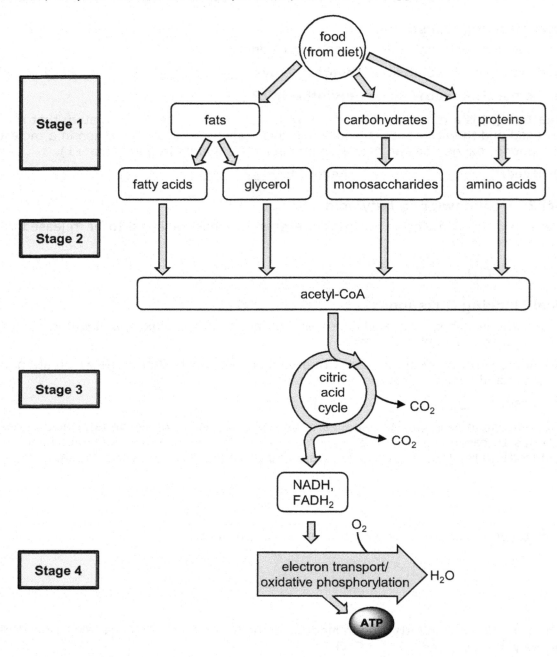

Critical Thinking Questions:

10. List the component molecules from food that remain at the end of stage 1 of metabolism.

11. Work as a team to describe the overall goal of stage 1 of metabolism in a sentence.

12. a. Breakdown of all three classes of food molecules results in a common intermediate at the end of stage 2. What is the common molecule?

    b. Discuss as a team and propose one or two advantages of having the same product at the end of stage 2.

13. a. What are the products of the citric acid cycle in stage 3?

    b. What product(s) feed into the next stage?

14. What are the products of electron transport and oxidative phosphorylation in stage 4?

15. Consider reactions (1) and (2) in Models 1 and 2.
    a. Do these reactions represent catabolism or anaboism? _____
    b. In the process you listed in part (a), which atom is oxidized? Which is reduced?

    c. Other than water, what is the final product after all the stages of catabolism have been completed? (We will investigate this molecule further in CA 49B.)

    d. For each reactant and each product in reactions (1) and (2), state the stage of catabolism in Model 2 in which it is found.

16. What is(are) the major ideas(s) of this activity?

Table 1: Enzymes that catalyze the initial hydrolysis of food molecules, and their locations and products

Substrate	Enzyme (Location)	Products
complex carbohydrates	amylase (saliva) glycosidases (small intestine)	simple sugars and disaccharides
proteins	peptidases (stomach) proteases (small intestine)	amino acids and short peptides
lipids	lipases (small intestine)	fatty acids and monoacylglycerols

Exercises:

1. Consider Table 1 above. What type(s) of enzymes would be involved in beginning digestion (Stage 1) of the following food sources? Where in the digestive tract would this occur?

   a. Potato starch *digestion begins in the oral cavity by amylases*

   b. Corn oil *digestion begins in the small intestines by lipases*

   c. Soy protein *digestion begins in the stomach by peptidases*

   d. Shortening from a pie crust *digestion begins in the small intestines by lipases*

2. a. Using the information from reaction (1) in Model 1, determine the amount of energy released for one mole of glucose in kcal/g. Assume that glucose has a molecular weight of 180 g/mol. *233.6 kcal*

   b. Using the information from CTQ 8, determine the amount of energy released for one mole of palmitic acid ($C_{16}H_{32}O_2$). Assume that palmitic acid has a molecular weight of 256 g/mol. *773.8 kcal*

3. Examine the nutritional label from a box of breakfast cereal shown below:

   a. How many grams of carbohydrates are in one serving of this cereal? *38g*

   b. How many Calories (nutritional Calories are equal to kcal) are derived from the carbohydrates in one serving of the cereal? Assume that the energy from glucose calculated in Exercise 1 (a) approximates the energy from all carbohydrates. For simplification, round the energy value to one significant figure.

   137.2 kcal

Nutrition Facts	
Serving Size 3/4 cup (55 g)	
Serving Per Container 5	
Amount Per Serving	
Calories 225 Calories from Fat 45	
	% Daily Values*
Total Fat 5g	8%
Saturated Fat 0.5g	3%
Trans Fat 0g	
Sodium 135mg	6%
Total Carbohydrate 38g	13%
Dietary Fiber 10g	40%
Sugars 8g	
Protein 7g	14%
*Percent Daily Values are based on a 2,000 calorie diet.	

c. How many Calories are derived from fat in one serving of this cereal? (Assume that the reaction in Model 2 can be used as an approximation for the energy derived from fats.) For simplification, round the energy value to one significant figure.

45

d. How many Calories are derived from protein in one serving of this cereal? (Recall: total Calories are comprised of carbohydrates, fat, and protein).

7g

e. How much energy would be released per gram of protein? (Again, round to one significant figure)

28 KCAl

4. Propose at least two reasons why we need energy from food to survive.

To keep the body breathing and for growth and repair of tissues

5. If we don't eat, how might we still have energy to survive for a time?

The body will take energy from fat in the body

6. Read the assigned pages in your textbook and work the assigned problems.

Metabolic Energy
(What are the energy currencies of the cell?)

Information:

The energy produced in catabolic reactions is stored in many different molecules, but the most important of these is ATP, **a**denosine **tri**phosphate. It is said that an organism must produce and use its weight in ATP each day.

Figure 1: "High-energy" phosphate bonds (indicated in bold) in ATP can be hydrolyzed to release energy.

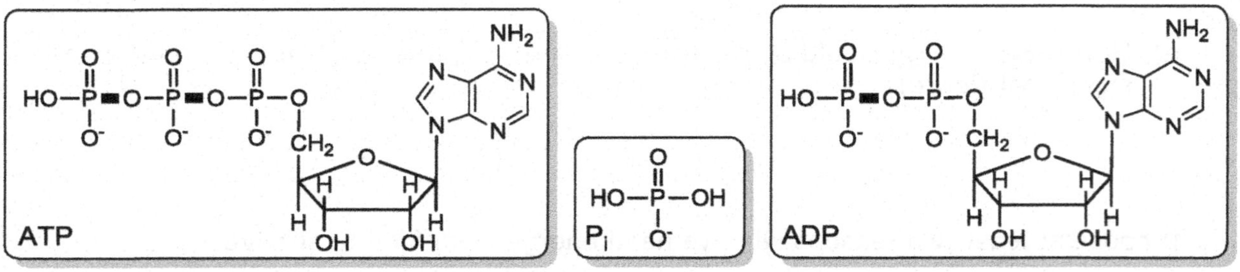

$$ATP + H_2O \rightleftharpoons P_i + ADP + 7.3 \text{ kcal/mol}$$

Note the abbreviations in Figure 1 and what they mean: ATP is an adenosine with three phosphates connected together *via* phosphoanhydride bonds; ADP is an adenosine with two phosphates connected together *via* a phosphoanhydride bond; and P_i means *inorganic phosphate*, shown as dihydrogenphosphate ($H_2PO_3^-$). You will sometimes see ⓟ (a P in a circle) as a shortcut to designate a phosphate group.

It is important to note that energy is **not** released by **breaking** the P-O bond, but by the **making** of the new O-H and P-O bonds during hydrolysis.

Critical Thinking Questions:

1. The high energy phosphate bond in ADP can be hydrolyzed to release another 7.3 kcal/mol of energy, as shown:

    $$ADP + H_2O \rightleftharpoons AMP + P_i + 7.3 \text{ kcal/mol}$$

    By analogy to Figure 1, draw chemical structures for the two products of this hydrolysis.

2. If an ATP molecule is hydrolyzed (by two water molecules) all the way to AMP (and two P_i) in two steps, how many total kcal/mol of energy could be released? Explain.

3. Work with your team to determine how much energy would be required to synthesize ATP from ADP. Be prepared to share your answer with the class.

Model: ATP is the energy currency of the cell

Consider the phosphorylation of glucose reaction below. The reaction is unfavorable (endothermic) because 3.3 kcal/mol of energy are required.

$$\text{Glucose} + P_i + 3.3 \text{ kcal/mol} \rightleftharpoons \text{glucose-6-phosphate} + H_2O \qquad (1)$$

If ATP is present, it can be hydrolyzed to both provide the needed P_i and the energy.

$$\text{ATP} + H_2O \rightleftharpoons \text{ADP} + P_i + 7.3 \text{ kcal/mol} \qquad (2)$$

Critical Thinking Questions:

4. Copy **all** of the reactants from **both** reactions (1) and (2) in the Model into the box below. Repeat for the products.

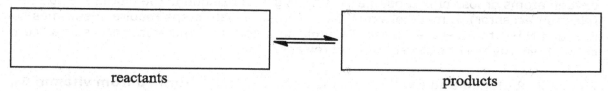

reactants   products

5. Cross out any chemicals that are present in both the reactants and the products.

6. Rewrite the net reaction from CTQ 4 in the space below, **omitting** the energies (kcal values).

7. What is the **net** energy required or released by the combined reaction in CTQ 6? Add the net amount of kcal/mol to either the reactant or product side, as appropriate.

8. The net reaction in CTQ 6 is the result when reactions (1) and (2) are **coupled**.

   a. Does the net reaction **require** energy, or is energy **released** (circle one)?

   b. Is the net reaction **exothermic** or **endothermic** (circle one)?

   c. Is the net reaction **favorable** or **unfavorable** (circle one)?

9. Consider the reaction below:

   $$\text{fructose-6-phosphate} + P_i + 3.4 \text{ kcal/mol} \rightleftharpoons \text{fructose-1,6-bisphosphate} + H_2O \quad (3)$$

   a. In this reaction, is energy **required**, or is energy **released** (circle one)?

   b. If energy from ATP is required for this reaction, in order to provide sufficient energy, would the product of ATP hydrolysis need to be **ADP** or **AMP** (circle one)? Explain your choice. Be prepared to share your answer with the class.

10. Consider the reaction below:

    $$\text{phosphoenolpyruvate} + H_2O \rightleftharpoons \text{pyruvate} + P_i + 14.8 \text{ kcal/mol} \qquad (4)$$

    a. In this reaction, is energy **required**, or is energy **released** (circle one)?

    b. Would energy from ATP be required for this reaction? _____

    c. Is sufficient energy produced by this reaction to make ATP from ADP + P_i? Explain.

CA49B

11. ATP hydrolysis, as shown in reaction (2), has an energy change larger than that of reaction (3), but smaller than that of reaction (4). Work with your team to propose a reason why this is important. Be prepared to share your answer with the class.

Information: Other energy currencies

We have defined oxidation as the loss of electrons, and we recognize that the addition of oxygen atoms or loss of hydrogen atoms are signs of oxidation. The ultimate oxidizing agent (electron acceptor) in metabolism is O_2, but intermediate steps require coenzymes such as FAD and NAD^+ to accept electrons. Each of these cofactors therefore serves as a "currency" in which energy may be stored for later release.

Figure 2: Reduced and oxidized forms of NAD^+. NAD^+ (formed from vitamin B_3, niacin) is involved in oxidations producing C=O double bonds.

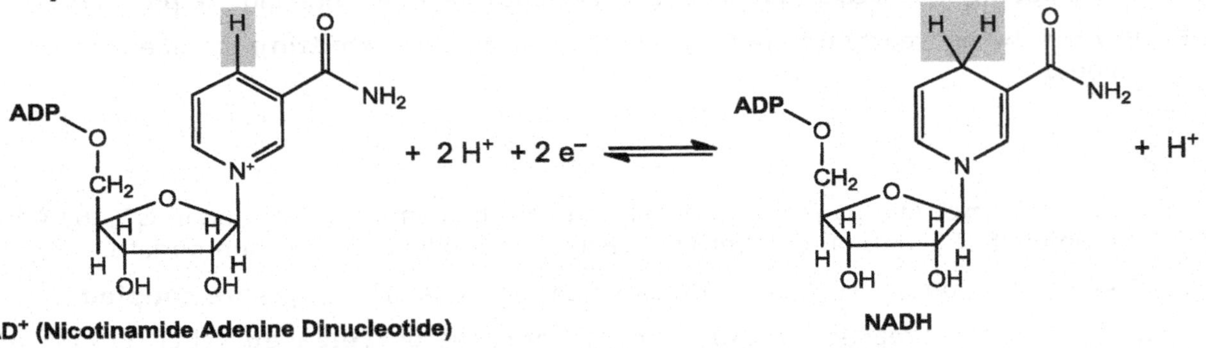

H
O
NH₂
ADP
O
CH₂
O
N⁺
H
H
H
H
OH
OH
+ 2 H⁺ + 2 e⁻
H H
O
NH₂
ADP
O
CH₂
O
N
H
H
H
H
OH
OH
+ H⁺
NAD⁺ (Nicotinamide Adenine Dinucleotide)
NADH

For example, ethanol is oxidized to acetaldehyde in liver cells according to the reaction:

$$CH_3CH_2OH + NAD^+ \xrightarrow{\text{alcohol dehydrogenase}} CH_3CHO + NADH + H^+ \quad (5)$$

Critical Thinking Questions:

12. When a molecule accepts electrons, does it become reduced or oxidized? ____________

13. Which molecule in Figure 2 is in the reduced form: **NAD^+** or **NADH** (circle one)?

14. Consider Figure 2, including the example reaction (5) at the bottom.
   a. Why is the reaction of NAD^+ to NADH and H^+ called a **reduction** of NAD^+?

   b. How many electrons will NADH "carry" from the reaction? _____

   c. Due to the shorthand notation for the aldehyde product in reaction (5), the location of the C=O that is produced in the reaction is not obvious. Draw a chemical structure that shows the carbonyl group. Does your team agree?

Figure 3: Reduced and oxidized forms of FAD; ADP = adenosine diphosphate. FAD (formed from vitamin B₂, riboflavin) is involved in reactions that produce a C=C double bond.

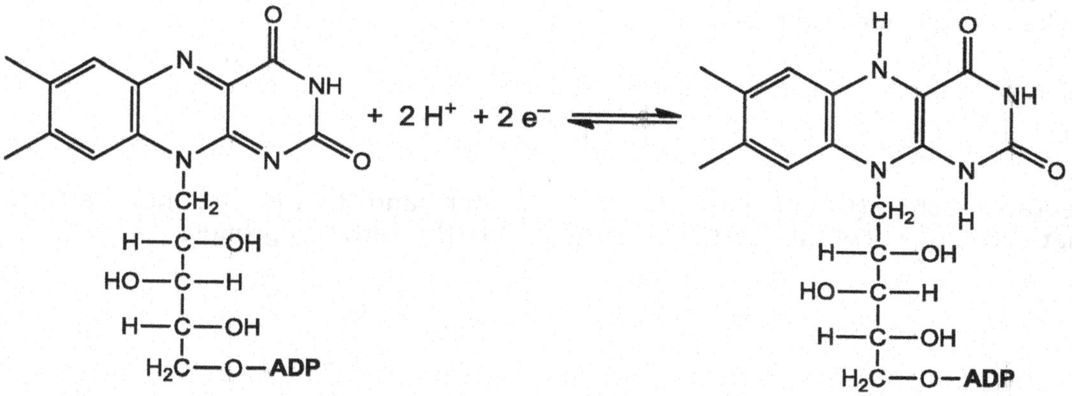

FAD (Flavin Adenine Dinucleotide)   FADH₂

For example, succinate is converted to fumarate by the following reaction:

$$^-OOC\text{-}CH_2\text{-}CH_2\text{-}COO^- + FAD \xrightarrow{\text{succinate dehydrogenase}} {}^-OOC\text{-}CH\text{=}CH\text{-}COO^- + FADH_2 \quad (6)$$

Critical Thinking Questions:

15. Where are the two hydrogen atoms added to FAD? Circle them in the structure of FADH₂ in Figure 3.

16. Which molecule in Figure 3 is in the oxidized form? _____

17. How many electrons will FADH₂ "carry" from the reaction?

18. What cofactor (NAD⁺, NADH, FAD, FADH₂) would be required in the following reaction? Work with your team to explain. Be prepared to share your answer with the class.

pyruvate   lactate dehydrogenase   lactate

19. Which is in the more oxidized form—pyruvate or lactate? Explain how you can tell. Be prepared to share your answer with the class.

20. Work with your team to list 2-3 major concepts from this activity. Be prepared to share your answer with the class.

21. List one strength of your teamwork today, and tell why it helped your learning.

Exercises:

1. Considering reactions (1) and (2) in the Model and your answers to CTQ 4–8, write a sentence or two to explain how coupling an unfavorable reaction with ATP hydrolysis can make the reaction favorable.

2. Redraw reaction (6) in Figure 3 without the shorthand (that is, use chemical structures that explicitly show the locations of the C=O in the carboxyl groups.

3. Draw a complete balanced reaction for the conversion of pyruvate to lactate (CTQ 18), showing all reactants, products, enzymes and cofactors.

4. We saw in this activity that oxidation reactions of alcohol groups in carbohydrates can provide enough energy to transfer electrons to NAD^+.

    a. Which form of this cofactor is the more oxidized form—NAD^+ or NADH?

    b. Which form is the more reduced form?

    c. Given that in an oxygen-containing environment, oxidation reactions are generally favorable (spontaneous), which form (**oxidized** or **reduced**) is at a higher potential energy level?

    d. In general, would *oxidized cofactors* or *reduced cofactors* provide a "stockpile" of energy? Explain.

5. The enzyme nucleoside *diphosphate kinase* catalyzes the following reaction, with an equilibrium constant, K_{eq}, equal to 1:

$$GTP + ADP \rightleftharpoons GDP + ATP$$

Propose an explanation for why ATP and GTP "store" equivalent amounts of energy.

6. Read the assigned pages in your textbook and work the assigned problems.

Digestion
(How is food absorbed?)

Model 1: Digestion in the body

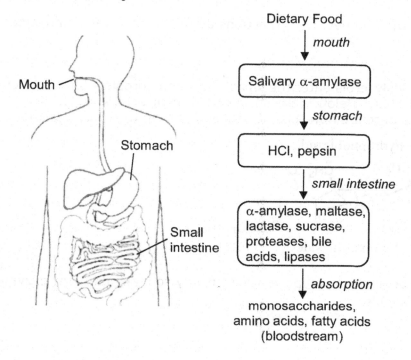

Critical Thinking Questions:

1. Which organs in the body are important for the digestion of food?

2. Where in the body does absorption of small molecules take place?

3. a. Give 2 or 3 examples of complex carbohydrates that would be found in the diet.

    b. To be absorbed and reach the bloodstream, to what form must they be converted?

4. Recall that enzymes tend to be named after the substrates or reactions they catalyze. Discuss with your team and determine which macromolecule undergoes initial hydrolysis in the mouth by α-amylase.

5. Pepsin is originally formed as a **zymogen**, an inactive enzyme precursor, and is activated in a low pH environment.

    a. Describe the effect a low pH environment can have on protein structure.

    b. How might this environment affect typical dietary proteins?

c. Why would the effect you described in part (b) be an important part of digestion?

6. Pepsin is an enzyme which cleaves specific peptide bonds. Which class of food macromolecule would undergo initial hydrolysis by pepsin in the stomach?

7. Based upon their names, what reactions do maltase, sucrase, and lactase catalyze?

8. Dietary triacylglycerols enter the small intestine in droplets too large for enzymes to interact with them. Below is the structure of cholic acid, a major bile acid. Work with your team to describe how this molecule could help with lipid digestion.

hydrophobic side

hydrophilic side

9. Work with your team to propose a reason why so many different enzymes are needed in the small intestine.

10. The proteases found in the small intestine are originally formed in the pancreas as zymogens. Propose a reason that zymogens would be important under these circumstances.

11. Incomplete digestion products of triacylglycerols, mainly diacylglycerols and monoacylglycerols, require special transport across the small intestine membranes and within the blood. Work with your team to describe why these molecules would need special transport in the blood (in micelle type structures) and amino acids and monosaccharides would not.

12. Summarize where each macromolecule is hydrolyzed to smaller components within the body.

13. Indicate how each team member performed their role today and any area(s) for improvement.

Exercises:

1. What are the primary monosaccharides produced from our diets (hint: consider your answer to CTQ 7)?

2. What type(s) of enzymes would be involved in beginning digestion of the following food sources? Where in the digestive tract would this occur?

    a. Corn starch

    b. Fish oil

    c. Protein from a tuna steak

3. A student had milk with her cereal this morning. Describe how the lactose would be digested and absorbed. What would happen to the fat? The proteins?

4. Although no digestion of triacylglycerols occurs in the stomach, the heat and churning action breaks up fats into smaller droplets for the intestines. This process takes longer than for carbohydrates and proteins and slows down the rate at which food leaves the stomach. Explain why you might get hungry again soon after eating a low-fat, high-carbohydrate meal.

5. What is the role of bile acids in the digestion of fats?

6. Lipases catalyze the hydrolysis of triacylglycerols. What would be the products of the following reaction?

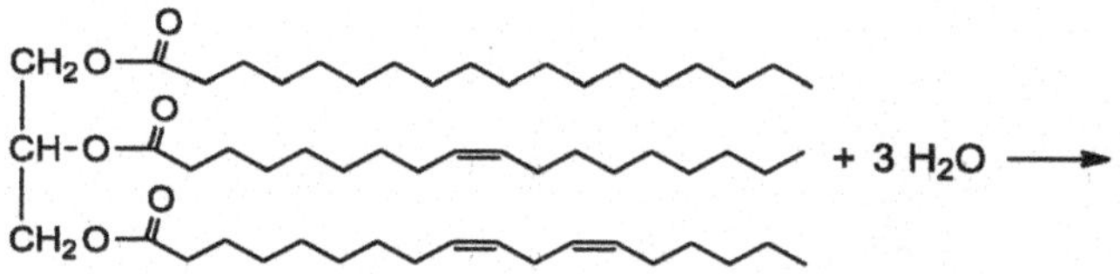

a fat molecule

7. Read the assigned pages in the text and work the assigned problems.

(This page is intentionally left blank.)

Glycolyis
(How is glucose converted to pyruvate?)

Model 1: The three major fates of glucose

Catabolism is the breakdown of food molecules for energy. Initial stages of catabolism including glycolysis occur in the cytosol (cytoplasm), while later stages (and most of the oxidation reactions) occur in mitochondria.

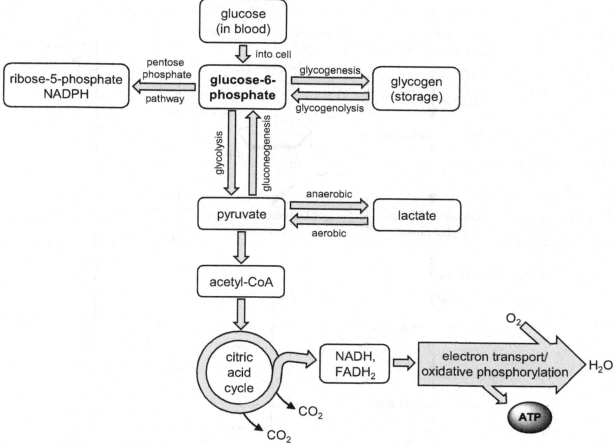

Critical Thinking Questions:

1. Where in the cell does glycolysis occur? _____

2. What are the three pathway options for glucose-6-phosphate?

3. a. When energy levels are abundant, which of the three pathways would glucose follow? Why?

   b. When energy is needed, which of the three pathways would glucose follow? Why?

   c. Discuss with your team and propose a situation when glucose would be likely to follow the third pathway. (Hint: consider the products that are made and when and where they would be needed.)

Model 2: The ten steps of glycolysis

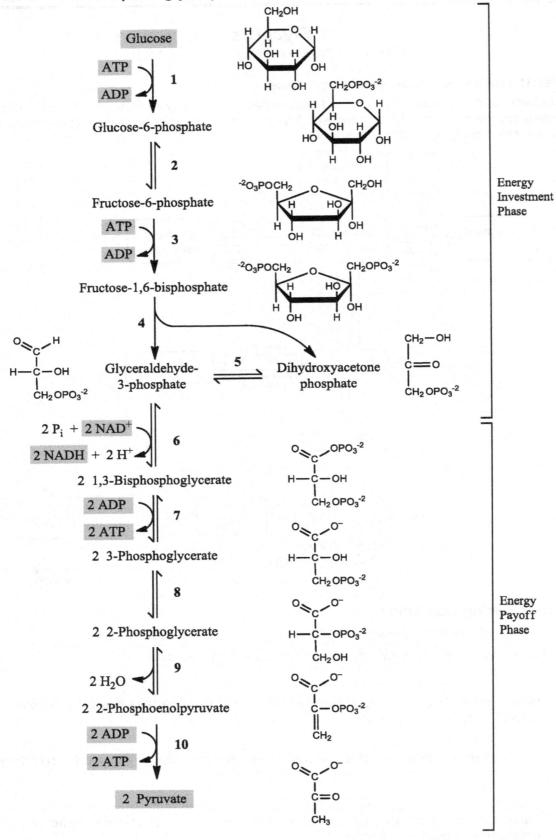

Energy Investment Phase

Energy Payoff Phase

Critical Thinking Questions

Refer to Model 2 to help you answer the following questions.

4. Which steps of glycolysis are referred to as the "energy investment phase?"

5. Work with your team to explain **why** these steps would be called the "energy investment phase."

6. In step 4 of glycolysis, the 6-carbon molecule fructose-1,6-bisphosphate is split into two 3-carbon molecules—glyceraldehyde-3-phosphate and dihydroxyacetone phosphate. What must happen to the dihydroxyacetone phosphate in order for it to continue proceeding through the glycolytic pathway?

7. Consider steps 6-10 of glycolysis.
    a. Why is there a "2" in front of each substrate in steps 6-10 (e. g., 2 1,3-Bisphospho-glycerate, 2 3-Phosphoglycerate, *etc.*)

    b. Why are these steps referred to as the "energy payoff phase?"

8. ATP is used in two steps (1 and 3) and generated in two steps (7 and 10). Explain, then, how glycolysis gives a net yield of 2 molecules of ATP.

9. Only one of the ten steps of glycolysis is an oxidation. Which one? How can you tell? (Hint: look at the cofactors required in each step.) Does your team agree?

10. Work with your team to determine which of the six classes of enzymes would be required for the following steps in Model 2.

    a. Step 3 _____

    b. Step 6 _____

    c. Step 2 _____

    d. Step 8 _____

11. Is oxygen (O_2) required as a substrate for any of the ten steps of glycolysis? _____

12. Based on your answer to CTQ 11, would glycolysis be considered an **aerobic** or an **anaerobic** pathway (circle one)?

13. Erythrocytes (red blood cells) contain no mitochondria. Can erythrocytes obtain energy from glycolysis? Work with your team to explain why or why not.

Model 3: Aerobic and anaerobic fates of pyruvate

Glycolysis does not require oxygen, but it does require an oxidizing agent, NAD^+, in step 6. Under aerobic conditions, the NADH produced during glycolysis is converted back to NAD^+ in the mitochondria, and glycolysis can continue. But under anaerobic conditions, the cell needs another method to regenerate the NAD^+, or else energy production would cease. This method in most organisms is to produce lactate (lactic acid), as shown in Figure 1.

Figure 1: Under anaerobic conditions, pyruvate is converted to lactate. The production of acetyl-SCoA under aerobic conditions is often called the "bridge" or "transition" step to the citric acid cycle.

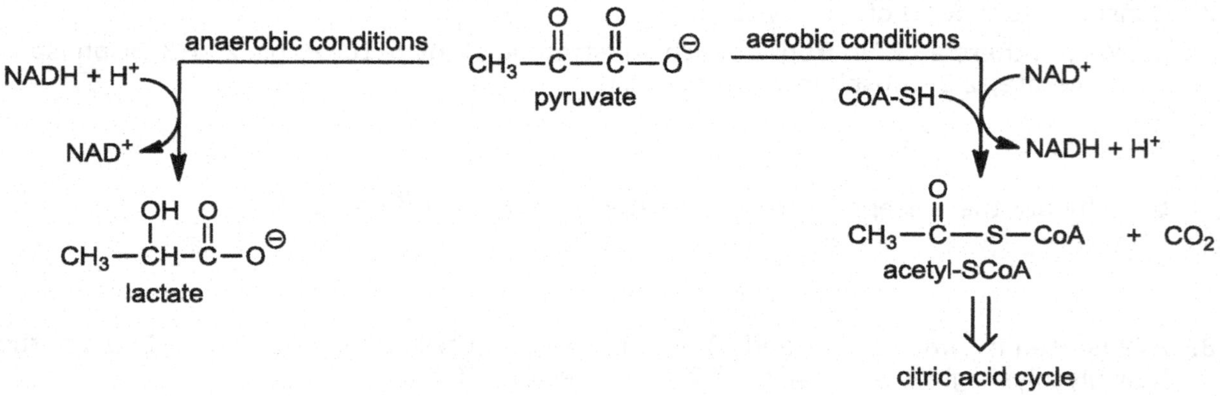

Critical Thinking Questions:

14. Why does the body need an option for converting pyruvate to lactate? (Hint: What else is produced in the reaction?)

15. Could glycolysis continue under anaerobic conditions if lactate could not be produced? Explain.

16. What is the coenzyme produced in the **aerobic** pathway in Figure 1? _____

17. a. Review Model 1. What pathway other than glycolysis generates ATP? _____

    b. Is oxygen needed for this process? _____

    c. Discuss with your team why an anaerobic option is critical in the body. Write your consensus answer below.

18. Lactic acid is, of course, acidic. When muscles contract, they squeeze the blood out of the vessels, and therefore are operating anaerobically. Work with your team to determine what happens to the pH of muscle cells during prolonged contraction. How might this help explain muscle soreness from a workout?

19. List one question remaining with your team from this activity.

20. What is one discovery or insight about glycolysis that you had today?

Exercises

1. When glucose enters the cell, it is immediately phosphorylated to make glucose-6-phosphate. Propose a reason why this step would be important.

2. Lactate eventually makes its way to the liver, where it is converted back into pyruvate. Explain why lactate production is called a "dead-end" pathway.

3. In the absence of oxygen, yeast cells regenerate NAD^+ by converting pyruvate into ethanol and carbon dioxide. Write a balanced chemical equation for this reaction.

4. The net chemical equation for the first step of glycolysis is:

$$\text{Glucose} + \text{ATP} \rightleftharpoons \text{glucose-6-phosphate} + \text{ADP}$$

Write a similar equation for the following steps of glycolysis:

a. Step 7:

b. Step 4:

c. Step 10:

d. Step 6:

e. Step 3:

5. Using only the information in Figure 1, it is possible to determine the class of enzyme used for each of the ten steps of glycolysis. Which of the six classes of enzymes would be required for the following steps?

   a. Step 1 _____

   b. Step 4 _____

   c. Step 5 _____

   d. Step 9 _____

6. Step one of glycolysis is irreversible in muscle cells, meaning that glucose-6-phosphate is not easily converted back to glucose. Why would this beneficial to muscle cells?

7. (optional) Glucose-6-phosphate is an allosteric inhibitor of hexokinase, the enzyme that catalyzes reaction 1 of glycolysis. Review the section in your text about enzyme inhibitors.

   a. How does an allosteric inhibitor work?

   b. Why would this enzyme be a good target for regulation by inhibition?

8. Read the assigned pages in your textbook and work the assigned problems.

(This page is intentionally left blank.)

Citric Acid Cycle
(How is pyruvate oxidized to CO₂?)

Model: The citric acid cycle oxidizes acetyl-CoA to CO₂ in eight steps

Recall that the end product of aerobic glycolysis is pyruvate, which is then converted to acetyl-CoA (See Figure 2 of ChemActivity 50). The citric acid cycle (CAC) takes the carbons of acetyl-CoA and oxidizes them to CO_2, producing the reduced cofactors NADH and FADH₂, and a small amount of ATP.

Seven of the eight CAC enzymes are found in the mitochondrial matrix. The eighth is bound in the inner mitochondrial membrane.

Critical Thinking Questions:

Refer to the Model to help you answer CTQs 1-10.

1. Which metabolite (substrate) provides the source of carbons for the CAC to oxidize? How many carbons are provided?

2. In step 1 of the CAC, the acetyl group is transferred from acetyl-CoA to oxaloacetate. Step 8 produces oxaloacetate. Explain why this pathway is said to be a **cyclic** pathway.

3. a. In each turn of the CAC, two molecules of CO_2 are released. In which steps is CO_2 released?

   b. What do you think the dashed lines around the carboxyl groups in isocitrate and α-ketoglutarate indicate?

4. a. Four of the eight steps in the CAC are oxidation/reductions. Which four?

   b. Consider reaction 6, in which succinate is converted to fumarate. Work with your team to determine which molecule is oxidized and which molecule is reduced. How can you tell?

   c. Consider reaction 8. Which molecule is oxidized and which is reduced? How can you tell?

5. Recall the oxidations of alcohols (ChemActivity 37).

   a. Can a secondary alcohol be oxidized?_____ If so, what functional group is produced?

   b. Can a tertiary alcohol be oxidized?_____ If so, what functional group is produced?

   c. What do you think is the purpose of Step 2 of the CAC? Explain.

6. No ATP is produced directly by the CAC, but one GTP (guanosine triphosphate) is produced in Step 5, and this is equivalent in energy to ATP. Other "energy-currency" cofactors produced in the CAC are NADH and $FADH_2$. For each turn (8 steps) of the CAC, circle the number of each that is produced from **one** molecule of acetyl-CoA.

   Number of NADH produced in one turn of the CAC (circle one) –   **1   2   3**

   Number of $FADH_2$ produced in one turn of the CAC (circle one) –   **1   2   3**

7. Adding your results from CTQ 6 to the cofactors produced in glycolysis and the bridge step (see ChemActivity 50, Models 2 and 3), add up the number of ATP molecules and other cofactors produced in the entire oxidation of <u>one molecule of glucose</u> to CO_2 and H_2O *via* <u>glycolysis and the CAC</u>. Do your team's numbers match those of other teams?

    ATP _____   GTP _____   NADH _____ $FADH_2$ _____

8. a. What cofactors are needed as reactants for the four oxidation steps in the CAC?

    b. As the CAC proceeds, are those cofactors **oxidized** or **reduced** (circle one)? How can you tell?

    c. Suppose that there is an adequate supply of acetyl-CoA. In order for the CAC to continue to operate, what will have to happen to the cofactors produced in the four oxidation steps?

9. a. Work with your team to identify and place a box around the two carbon atoms from the initial acetyl-CoA molecule as they move through the first four steps of the cycle. (Hint: they are the top two in citrate)

    b. Are the two carbons from acetyl-CoA still present in oxaloacetate? _____

    c. Work with your team to explain the following statement: The carbons from acetyl-CoA are only oxidized to CO_2 after two or more turns of the citric acid cycle.

10. The overall objective of metabolism is to generate energy in the form of ATP by oxidizing carbon and reducing oxygen. Discuss as a team, and write a complete sentence or two summarizing how the citric acid cycle has contributed to this objective.

11. List one strength of your teamwork today. Why was this beneficial?

12. What are two or three major concepts from this activity?

Exercises:

1. Using only the information in Model 1, work with your team to determine which of the six classes of enzymes would be required for the following steps?

   a. Step 4 _____

   b. Step 5 _____

   c. Step 6 _____

   d. Step 7 _____

2. The chemical equation for the first step of the CAC is:

   $$\text{oxaloacetate} + \text{acetyl-SCoA} + H_2O \rightleftharpoons \text{citrate} + \text{CoA-SH}$$

   Write a similar equation for the following steps of the CAC:

   a. Step 2:

   b. Step 4:

   c. Step 5:

   d. Step 7:

Figure 1: The chemical structure of Coenzyme A. Note the reactive thiol group.

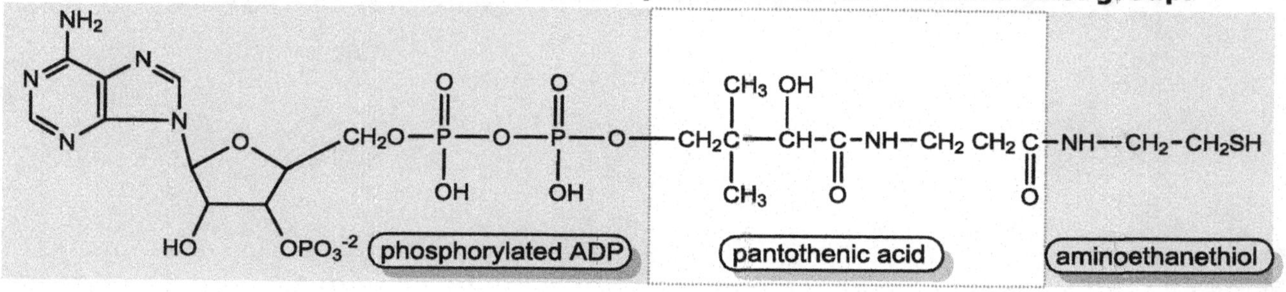

3. The part of Coenzyme A (see Figure 1 above) that originated in pantothenic acid is actually only an acid *residue*. It is the result of two condensation reactions—with aminoethanethiol on one end, and with the phosphorylated ADP on the other end. Draw the structure of pantothenic acid. (Recall that condensation reactions normally join two molecules and release a water molecule.)

Figure 2: Coenzyme A (CoA or CoA-SH) can carry an acetyl group. The resulting molecule is abbreviated "acetyl-CoA" or "acetyl-SCoA"

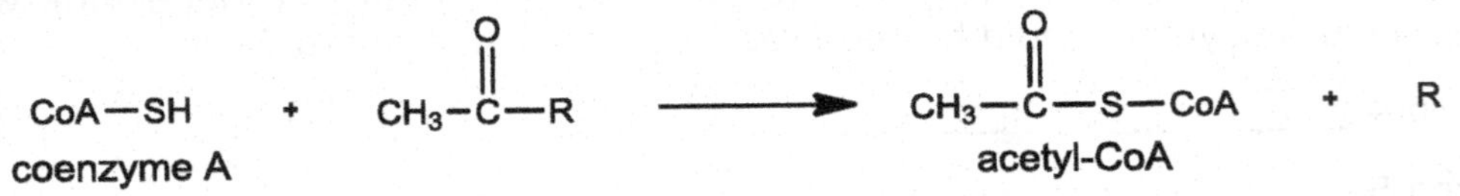

$$CoA-SH \quad + \quad CH_3-\overset{\overset{\displaystyle O}{\|}}{C}-R \quad \longrightarrow \quad CH_3-\overset{\overset{\displaystyle O}{\|}}{C}-S-CoA \quad + \quad R$$

coenzyme A   acetyl-CoA

4. Coenzyme A functions as a carrier of carbons from different sources. What is the reactive functional group in coenzyme A?

5. Using your textbook as a reference, sketch and label a diagram of a mitochondrion (in a human muscle cell, for example).

6. Write a general description of what cellular respiration is, including when it occurs.

7. Read the assigned pages in your text, and work the assigned problems.

Electron Transport/Oxidative Phosphorylation
(How can energy from the reduced cofactors be used to make ATP?)

At the end of the citric acid cycle, all six carbons of glucose have been oxidized to CO_2, and a few nucleotide triphosphates (ATP and GTP) have been produced. However, most of the energy has been saved in the reduced cofactors NADH and $FADH_2$.

The proteins and coenzymes in Models 1 and 2 carry electrons within and between the ETC enzyme complexes, either one or two at a time.

Model 1: Protein electron carriers in the Electron Transport Chain (ETC)

Figure 1: Iron-sulfur proteins are proteins containing *iron-sulfur clusters*—in which the iron is at least partially coordinated by sulfur. A typical iron-sulfur cluster is shown; curved lines indicate continuation of the protein backbone.

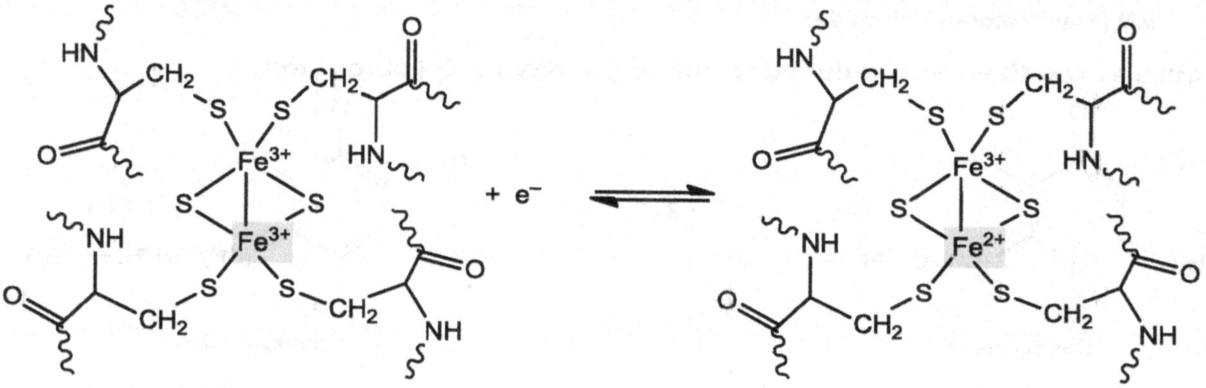

Figure 2: Oxidized and reduced forms of the heme group of a *b* cytochrome. Cytochromes are heme-containing proteins.

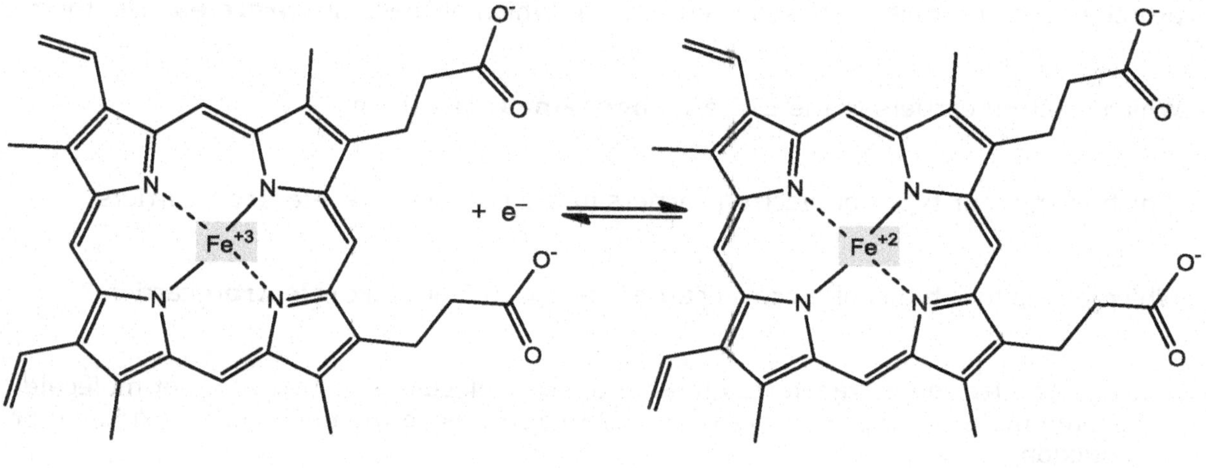

Critical Thinking Question:

1. When an atom or molecule loses electrons, is it **oxidized** or **reduced** (circle one)?

2. Consider the electron carriers in Model 1. Which form of iron – Fe^{3+} or Fe^{2+} – is more oxidized? Which is more reduced? Explain how you can tell.

Model 2: Coenzyme electron carriers in the ETC

Figure 3: The structure of FMN is similar to FAD (ChemActivity 49A), but with a phosphate group (at bottom) in place of ADP.

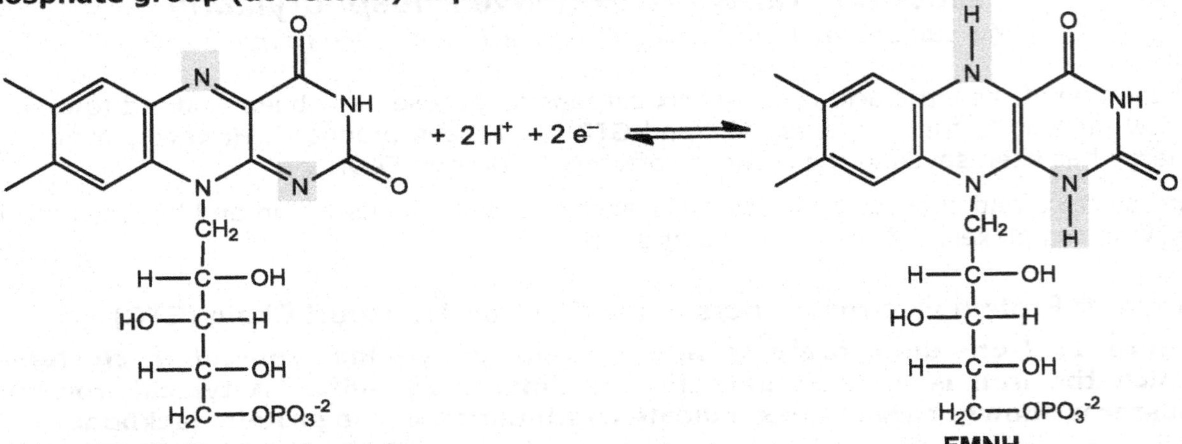

$+ 2 H^+ + 2 e^-$

FMN (Flavin Mononucleotide)

FMNH$_2$

Figure 4: Oxidized and reduced forms of Coenzyme Q (ubiquinone)

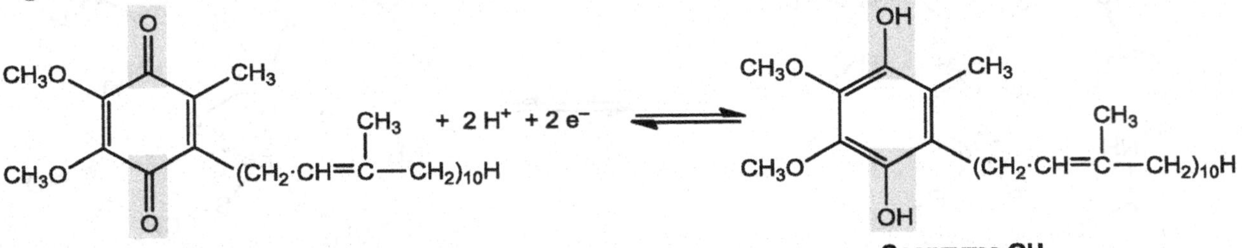

$+ 2 H^+ + 2 e^-$

Coenzyme Q

Coenzyme QH$_2$

Critical Thinking Questions:

Refer to Models 1-2 to help you answer CTQs 3-8.

3. Models 1 and 2 show structures of the four electron carriers in the ETC: iron-sulfur proteins, cytochromes, FMN and coenzyme Q. Which of these are **proteins**? List them.

4. Which electron carriers in the ETC are **coenzymes**? List them.

5. Which of the four types of electron carriers in the ETC are **one-electron** carriers?

6. Which of the four types of electron carriers in the ETC are **two-electron** carriers?

7. a. When an electron is transferred to an atom or molecule, does the atom or molecule become more oxidized or more reduced? Explain, using the definition of oxidation or reduction.

   b. Does the atom or molecule which donated the electron become more oxidized or reduced?

8. In Figure 3, does Coenzyme Q become more oxidized or reduced as it changes to Coenzyme QH_2?

9. Which biological pathways that we have studied produce NADH?

Model 2: The enzyme complexes of the ETC

The four enzyme complexes of the ETC (predictably named Complexes I – IV) are integral proteins in the inner mitochondrial membrane. They serve to take the electrons from the reduced cofactors NADH and $FADH_2$ and transfer them to O_2 to make H_2O.

Figure 5: Both the electron-transport chain (consisting of four enzyme complexes and other cofactors) and ATP synthase (Complex V) operate in the inner mitochondrial membrane.

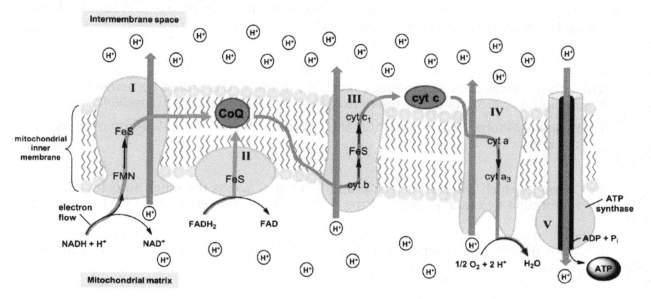

Critical Thinking Questions:

Refer to Figure 5 to help you answer all the remaining questions. The path of electrons through the various complexes and cofactors follows the shaded arrows.

10. Which complex accepts electrons from NADH?

11. Which complex accepts electrons from $FADH_2$?

12. Which complexes contain iron-sulfur clusters?

13. Which complexes contain cytochromes?

14. What molecule carries electrons from Complex I to Complex III? _____
    Does accepting electrons cause the **oxidation** or **reduction** of the molecule (circle one)?

15. a. Does NADH get oxidized or reduced in Complex I? Explain how you can tell.

    b. What pathway that we have studied will stop if NAD^+ is not regenerated?

16. For electrons entering the ETC from NADH, which of the complexes I – IV do the electrons pass through? List them.

17. For electrons entering the ETC from $FADH_2$, which of the complexes I – IV do the electrons pass through? List them.

18. What molecule carries electrons from Complex III to Complex IV? _____

19. What molecule is the final acceptor of electrons in complex IV? Work with your team to determine how this fits with the statement: The overall objective of metabolism is to generate energy in the form of ATP by oxidizing carbon and reducing oxygen.

20. If oxygen were unavailable, what would happen to the electron transport chain? What would happen to the citric acid cycle? Explain.

21. The energy released in all the redox reactions in the ETC is stored by active transport ("pumping") of protons out of the matrix to create a concentration gradient.

    a. Examine Figure 5 again. Which of the enzyme complexes I – IV pump protons out of the matrix?

    b. Through which complex do the protons pumped across the inner mitochondrial membrane eventually return to the matrix? What is another name for this complex?

    c. What unfavorable reaction is "powered" by the favorable return of protons to the matrix?

22. Suppose that as two electrons proceed through the ETC as shown in Figure 5, Complexes I, III, and IV each pump **two** protons out of the matrix. Considering your answers to CTQs 16 and 17, how many total protons are pumped:

    a. For the two electrons originating in NADH?

    b. For the two electrons originating in $FADH_2$?

23. Considering your answer to CTQ 22, if the oxidation of one NADH to NAD^+ in the ETC leads to production of 3 ATP molecules *via* ATP synthase (also known as Complex V), how many ATP would be produced from oxidation of one $FADH_2$? Explain your answer.

Information :

At the end of electron transport, a potential energy difference exists across the inner mitochondrial membrane, which can be described variously as a difference in charge, pH, or hydrogen ion (proton) concentration. This energy difference drives ATP synthesis. When protons pass through the integral membrane channel found in the ATP synthase complex, energy is provided to synthesize ATP.

Critical Thinking Questions:

24. Complete the table below with the numbers of products that would be obtained in the various steps from the aerobic oxidation of glucose. Then total the ATP that one glucose is "worth."

	Glycolysis	"Bridge to CAC"	CAC	Total	# of ATP "worth"	
ATP (or GTP)	2	0	2	4	4	Total
NADH						
FADH$_2$						

25. List one question remaining with your team about the ETC or oxidative phosphorylation.

26. Describe one area in which your team could improve next time.

Exercises:

1. When the electron carrier in CTQ 18 reaches complex IV, it transfers the electrons. Is this carrier oxidized or reduced in Complex IV. Explain how you know this.

2. What is a coenzyme? Find a definition in your textbook, and write it in your own words.

3. Consider the structure of ubiquinone (CoQ) and its physical location(s) in the ETC. Propose a function of the long hydrocarbon chain attached to the ring structure.

4. Biochemists measure pH to determine the values given in CTQ 23 for the number of protons pumped out by each ETC enzyme complex. Explain how variations in these measurements would lead to a different answer to CTQ 24.

5. In Figure 1, the iron atoms are coordinated to four amino-acid residues in a protein. All four amino-acid residues are the same. Identify the amino acid.

6. Read the assigned pages in your text, and work the assigned problems.

Additional Carbohydrate Pathways
(In what other ways is energy produced or used?)

Model 1: Glycogen metabolism

Glycogen (also called animal starch) is a polymer of α-D-glucose connected by α-1,4 linkages with α-1,6 branches. When energy is plentiful, excess glucose is converted into glycogen, much of which is stored in the liver. When energy is needed, the glycogen can undergo phosphorolysis to produce glucose-6-phosphate. These processes are summarized in Figure 1.

Figure 1: Glucose is stored as glycogen (synthesis), and released as glucose-1-phosphate (phosphorolysis)

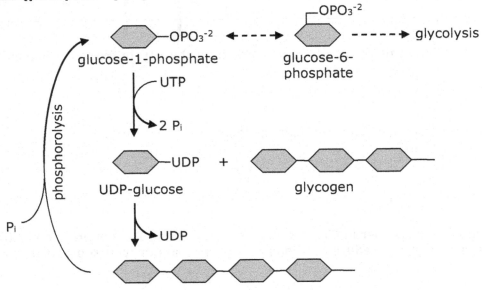

glycogen with one glucose added

Critical Thinking Questions:

1. a. How many "high-energy phosphate bonds" are hydrolyzed in order to convert glucose to glucose-6-phosphate? _____ (Refer to Model 2 of CA 50 if necessary.)

    b. Figure 1 above shows that glucose-6-phosphate can be converted to glucose-1-phosphate without using energy. How many "high-energy phosphate bonds" are hydrolyzed in order to attach one glucose-1-phosphate to glycogen? _____
    Explain how you can tell. Ensure that all team members agree with the explanation.

    c. In total, how many "high-energy phosphate bonds" are hydrolyzed in order to attach one glucose to glycogen? _____ Explain.

2. Explain why glycolysis starting from glycogen yields one more ATP than when starting with glucose itself.

Model 2: Gluconeogenesis

The brain normally requires glucose as an energy source (although it can survive on ketone bodies when fasting, if required). When glycogen supplies run low, liver enzymes can perform a process called gluconeogenesis, which literally means "new birth of glucose."

Figure 2: Gluconeogenesis produces glucose when glycogen stores are depleted. Pyruvate and oxaloacetate (from the CAC) are starting materials, but acetyl-CoA is not.

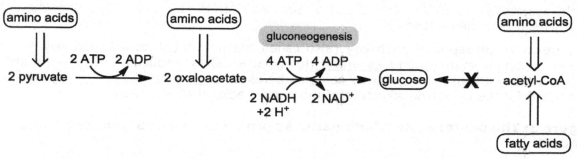

Critical Thinking Questions:

3. According to Figure 2, what small molecule results from breakdown of

   a. amino acids? _____

   b. fatty acids? _____

4. Model 2 shows that humans cannot convert acetyl-CoA into glucose.

   a. Refer to the Model of the Citric Acid Cycle in CA 51. To what molecule does acetyl-CoA attach in order to enter the cycle? _____

   b. What is the source of this molecule (see Figure 2 above)? _____

   c. Explain why it is probably unavoidable that persons on very low calorie diets often lose muscle mass along with body fat.

5. Humans can survive poorly on a diet of protein with very little fat or carbohydrates, but not at all on a diet of fat with little protein or carbohydrates. Discuss with your team, and write a consensus explanation.

Model 3: The pentose phosphate pathway is required for anabolic reactions

Catabolic pathways produce energy, and are considered oxidative, but anabolic (biosynthetic) pathways require energy and are considered reductive. While oxidative pathways require oxidized cofactors such as NAD^+ and FAD, reductive pathways require cofactors in their reduced form.

The electron transport chain keeps most of the cofactors NADH and $FADH_2$ in their oxidized forms, so they are not available in their reduced forms for biosynthesis. Therefore, reductive pathways make use of different cofactors, such as NADPH. (By attaching a phosphate group to NADH, it becomes NADPH.)

The **pentose phosphate pathway** (also called the phosphogluconate pathway or hexose monophosphate shunt) produces NADPH for what is called "reducing power"—the ability to perform biosynthesis. This pathway can also be used to produce ribose-5-phosphate, which is needed for the backbone structure of the nucleic acids RNA and DNA.

Figure 3: The pentose phosphate pathway produces ribose-5-phosphate and NADPH

$$3 \text{ glucose-6P} \xrightarrow[\substack{6\ NADP^+ \quad 6\ NADPH}]{\substack{3\ H_2O \quad 3\ H^+ + 3\ CO_2}} 3 \text{ ribulose-5P} \Longrightarrow \text{ribose-5P} \Longrightarrow \text{nucleic acids}$$

Critical Thinking Questions:

6. What are the two main products of the pentose phosphate pathway? For what purpose are the products used?

7. For each product you listed in CTQ 6, discuss with your team to propose a time or reason that the cell would need large amounts of that product and would therefore need to activate the pentose phosphate pathway.

Model 4: The three major uses of glucose

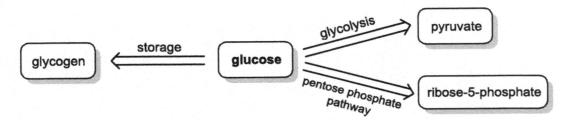

8. Of the three fates of glucose in Model 4 (storage, PPP, or glycolysis), two occur in most cells at all times. The other one alternates on and off. Explain. Ensure that all team members agree.

9. Which of the three pathways would increase in rate mainly during cell replication or division? Explain.

10. What questions remain with your team about the carbohydrate pathways that we have studied?

Exercises:

1. Glycolysis converts glucose to pyruvate, and gluconeogenesis converts pyruvate to glucose. The conversion of glucose → pyruvate → glucose by this path is called a futile cycle.

   a. Calculate the net number of ATP molecules that would be produced or required for this futile cycle.

   b. High blood glucose levels bring about the secretion of the hormone insulin. Insulin activates glycolysis and inactivates gluconeogenesis. Explain why it is beneficial for the organism to have both effects occur simultaneously.

   c. Would you expect insulin to activate or inactivate glycogen synthesis? Explain.

   d. Would you expect insulin to activate or inactivate glycogen phosphorolysis? Explain.

2. Read the assigned pages in your text, and work the assigned problems.

Fatty Acid Oxidation
(How is energy produced from fats?)

Model 1: Fatty acid activation

After fats (Figure 1) enter a cell, they are broken down to release individual (free) fatty acids (Figure 2). Fatty acids are activated for oxidation in the cytosol, in a process that requires ATP, by attaching them to acetyl-CoA to create fatty acyl-CoA molecules. The activated fatty acyl-CoA then enters the mitochondrial matrix for oxidation *via* a specific transport protein.

Figure 1: Stearin, a triacylglycerol

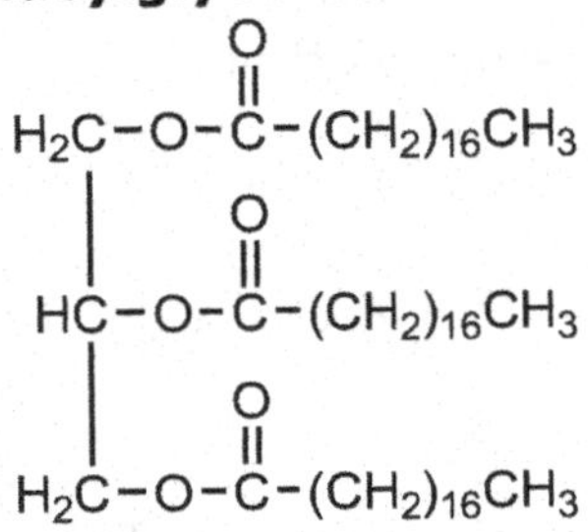

$$H_2C-O-\overset{\overset{O}{\|}}{C}-(CH_2)_{16}CH_3$$
$$HC-O-\overset{\overset{O}{\|}}{C}-(CH_2)_{16}CH_3$$
$$H_2C-O-\overset{\overset{O}{\|}}{C}-(CH_2)_{16}CH_3$$

Figure 2: Generic fatty acid structure with α (alpha) and β (beta) carbons labeled.

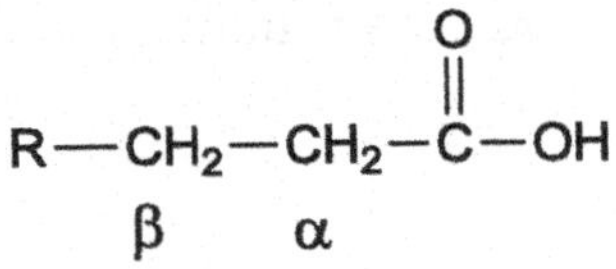

$$R-\underset{\beta}{CH_2}-\underset{\alpha}{CH_2}-\overset{\overset{O}{\|}}{C}-OH$$

Figure 3: Fatty acid activation takes place in the cytosol.

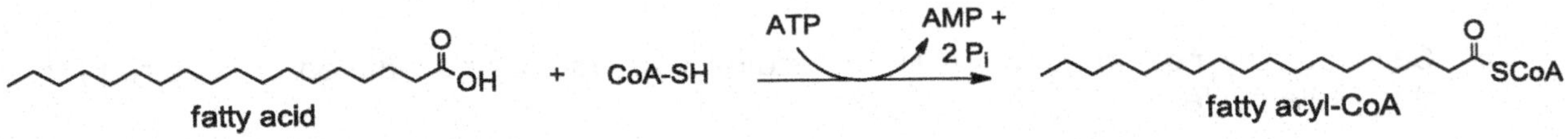

fatty acid   +   CoA-SH   →   (ATP → AMP + 2 P_i)   fatty acyl-CoA

Critical Thinking Questions:

1. How many free fatty acids are released from each fat molecule?

2. Where does fatty acid activation occur in the cell?

3. How many "high-energy phosphate bonds" are hydrolyzed during fatty acid activation (Figure 3)? Explain how can you tell, and ensure that all team members understand.

Model 2: β-oxidation removes 2-carbon units from a fatty-acyl-CoA to produce acetyl-CoA.

$$CH_3(CH_2)_{14} - \overset{\overset{\displaystyle H}{|}}{\underset{\underset{\displaystyle H}{|}}{C}}\beta - \overset{\overset{\displaystyle H}{|}}{\underset{\underset{\displaystyle H}{|}}{C}}\alpha - \overset{\overset{\displaystyle O}{||}}{C} - SCoA \qquad \text{fatty-acyl CoA}$$

FAD ⟶ FADH$_2$

$$CH_3(CH_2)_{14} - \overset{\overset{\displaystyle H}{|}}{C} = \overset{\overset{\displaystyle H}{|}}{\underset{\underset{\displaystyle H}{|}}{C}} - \overset{\overset{\displaystyle O}{||}}{C} - SCoA \qquad \textit{trans-}\Delta^2\textit{-enoyl-CoA}$$

H$_2$O

$$CH_3(CH_2)_{14} - \overset{\overset{\displaystyle OH}{|}}{\underset{\underset{\displaystyle H}{|}}{C}} - CH_2 - \overset{\overset{\displaystyle O}{||}}{C} - SCoA \qquad \beta\textit{-hydroxyacyl-CoA}$$

NAD$^+$ ⟶ NADH + H$^+$

$$CH_3(CH_2)_{14} - \overset{\overset{\displaystyle O}{||}}{C} - CH_2 - \overset{\overset{\displaystyle O}{||}}{C} - SCoA \qquad \beta\textit{-ketoacyl-CoA}$$

CoASH

$$CH_3(CH_2)_{14} - \overset{\overset{\displaystyle O}{||}}{C} - SCoA \quad + \quad CH_3 - \overset{\overset{\displaystyle O}{||}}{C} - SCoA$$

fatty acyl-CoA   acetyl-CoA
(2 C atoms shorter)

Critical Thinking Questions:

4. Where in the cell does β-oxidation of fatty acids occur? _____

5. Which of the four steps in β-oxidation are oxidation/reductions? How can you tell?

6. Which other pathways that we have studied produce the reduced coenzymes NADH and FADH$_2$?

7. Review the structures in Models 1 and 2, and propose a reason why this process is called β-oxidation.

8. Considering what you already know about metabolism, what will happen to the reduced coenzymes produced in β-oxidation?

9. Review Model 2. Which of the six classes of enzymes would be used in:

   a. Step 1 _____

   b. Step 2 _____

   c. Step 3 _____

   d. Step 4 _____

10. Suppose that step 1 of β-oxidation begins with an 18-carbon fatty-acyl-CoA, as in Model 2. Step 4 would then produce acetyl-CoA and a 16-carbon fatty-acyl-CoA.

   a. What would happen to the acetyl-CoA produced?

   b. What would happen next in order to further oxidize the 16-carbon fatty-acyl-CoA?

   c. Discuss with your team to come to an explanation of why β-oxidation is said to be a spiral rather than a cyclic pathway. Write the team explanation in a sentence.

11. Consider the β-oxidation of lauric acid, a 12 carbon fatty acid with a similar molecular weight to glucose.

   a. How many spirals of β-oxidation would be needed to completely degrade lauric acid to acetyl-CoA?

   b. Explain why the answer for (a) is not 6.

   c. How many $FADH_2$ molecules would be produced? _____

   d. How many NADH molecules would be produced? _____

   e. How many acetyl-CoA molecules would be produced? _____ What will happen to the acetyl-CoA molecules once they are formed?

   f. Remembering that each acetyl-CoA can be oxidized in the citric acid cycle to produce 3 NADH, 1 $FADH_2$, and 1 GTP, total the number of ATP produced from the C_{12} fatty acid after complete aerobic oxidation, including electron transport. Don't forget to subtract the number that you gave in the answer to CTQ 3. Show your work in the space below.

12. Compare the total number of ATP molecules generated from one molecule of lauric acid (12 carbons) to the total from two glucose molecules (12 carbons, CA 52). If you have approximately equal masses of fat or carbohydrates, which produces more energy?

13. Discuss with your team and propose an explanation for why organisms store excess energy as fat rather than as carbohydrates.

Model 3: Ketone bodies

When blood carbohydrate concentrations are low, such as during fasting, acetyl-CoA cannot attach to oxaloacetate to enter the citric acid cycle in order to be used for energy, and so the concentration of acetyl-CoA builds up. When this happens, other enzymes catalyze a condensation of two molecules of acetyl-CoA to form blood-soluble molecules called ketone bodies.

Figure 5: An excess of acetyl-CoA leads to the formation of the "ketone bodies" acetoacetic acid, β-hydroxybutyric acid, and acetone

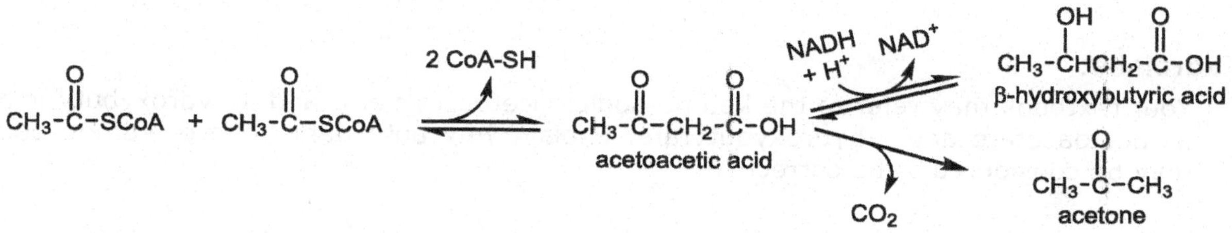

Critical Thinking Questions:

14. Locate and *circle* all the ketone groups in Figure 5. Check that all team members agree.

15. Locate and draw a *box* around all the carboxylic acid groups in Figure 5. Check that all team members agree. What affect would high levels of these molecules cause to the pH of the blood?

16. If the concentration of ketone bodies becomes too high, as can happen in diabetics or to those on a very low carbohydrate diet, a condition called **ketoacidosis** can result. This condition affects the ability of the hemoglobin to carry oxygen, and breathing can become difficult. Untreated, this condition can lead to coma, or even death. Considering your answers to CTQs 14 and 15, work with your team to develop a hypothesis on why this condition is called ketoacidosis.

17. When concentrations of ketone bodies are high, acetone can diffuse out of the bloodstream into air in the lungs. Why might a physician sniff the breath of a patient on a "low-carb" diet to check his health?

18. Ketone bodies can be removed by the kidneys and excreted in the urine. Would this increase or decrease the effectiveness of a low carbohydrate diet? Explain your answer.

19. What are the major concepts in this activity?

20. What is one area in which your team can improve next time? How can you achieve this?

Exercises:

1. Your textbook may refer to the ketone bodies acetoacetic acid and β-hydroxybutyric acid as acetoacetate and β-hydroxybutyrate. Explain why both definitions of ketone bodies may be considered to be correct.

2. Consider stearic acid, an 18-carbon saturated fatty acid.
   a. Calculate the total number of ATP that could be produced by the complete oxidation of stearic acid.

   b. Calculate the ratio of the number of ATP formed per carbon atom in stearic acid.

   c. Considering the complete oxidation of glucose to yield 38 ATP, calculate the ratio of the number of ATP formed per carbon atom in glucose.

3. Carbons that are in more reduced oxidation states can release more energy when they are oxidized. Considering your answers to Exercise 2, are the carbons in a fatty acid, on average, more <u>reduced</u> or more <u>oxidized</u> than those in a carbohydrate? Explain.

4. Read the assigned pages in your text, and work the assigned problems.

(This page is intentionally left blank.)

Fatty Acid Synthesis
(How is fat made?)

Model 1: Fatty acids are synthesized from acetyl CoA

Fatty acid synthesis is a spiral pathway similar to the reverse of β-oxidation. The main differences are the location (synthesis takes place in the cytosol) and the coenzymes and cofactors used. Also, the fatty-acyl groups are bonded to an acyl carrier protein (ACP) instead of to coenzyme A, though the attachment is a thioester bond in both cases.

Initial Steps:

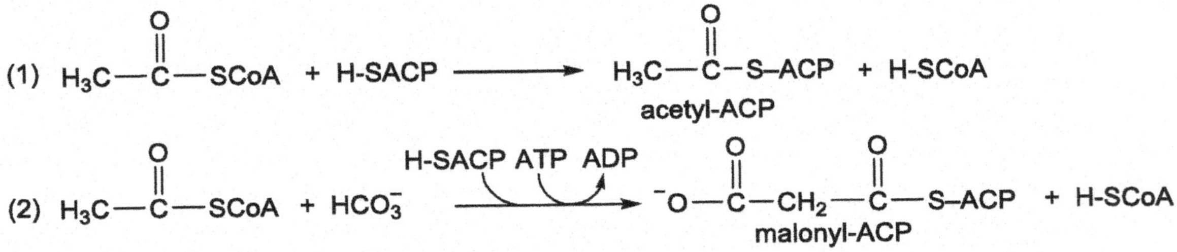

(1) $H_3C—\overset{\overset{O}{\|}}{C}—SCoA$ + H-SACP $\longrightarrow$ $H_3C—\overset{\overset{O}{\|}}{C}—S–ACP$ + H-SCoA
acetyl-ACP

(2) $H_3C—\overset{\overset{O}{\|}}{C}—SCoA$ + HCO_3^- $\xrightarrow{\text{H-SACP ATP ADP}}$ $^-O—\overset{\overset{O}{\|}}{C}—CH_2—\overset{\overset{O}{\|}}{C}—S–ACP$ + H-SCoA
malonyl-ACP

Elongation:

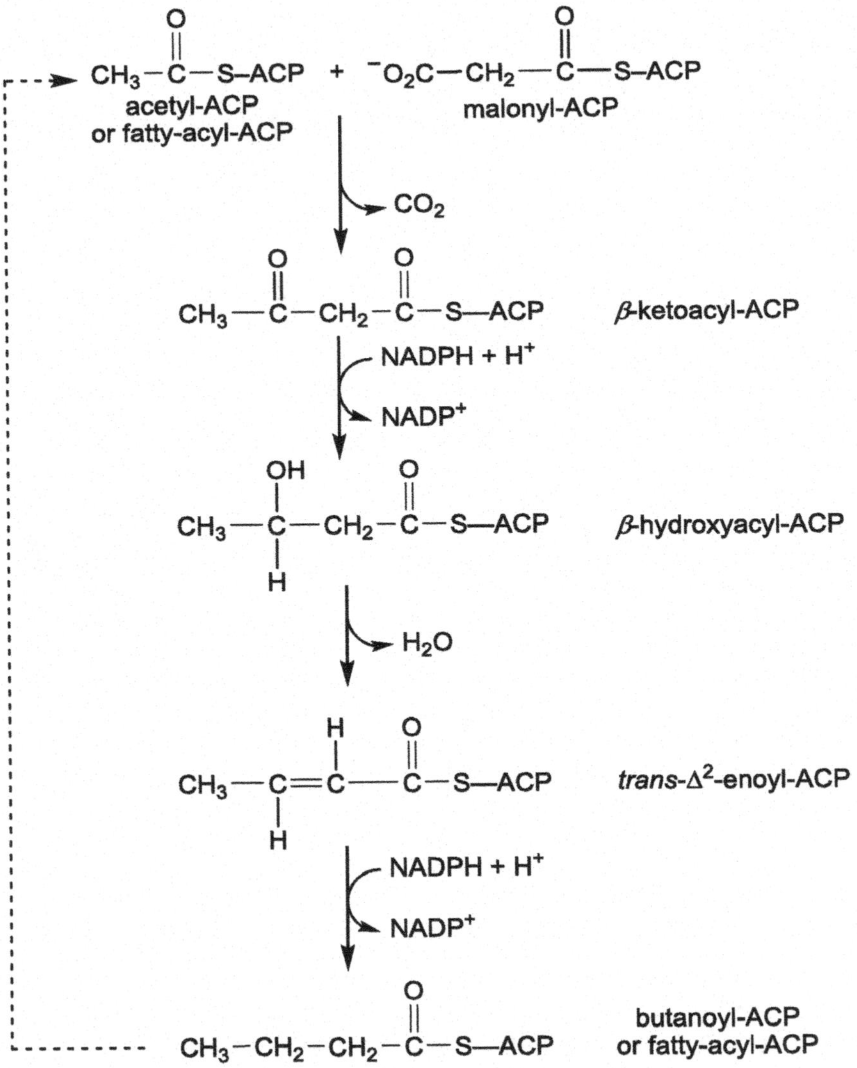

$CH_3—\overset{\overset{O}{\|}}{C}—S–ACP$ + $^-O_2C—CH_2—\overset{\overset{O}{\|}}{C}—S–ACP$
acetyl-ACP   malonyl-ACP
or fatty-acyl-ACP

$\downarrow$ CO_2

$CH_3—\overset{\overset{O}{\|}}{C}—CH_2—\overset{\overset{O}{\|}}{C}—S–ACP$ β-ketoacyl-ACP

$\downarrow$ NADPH + H$^+$
$\downarrow$ NADP$^+$

$CH_3—\overset{\overset{OH}{|}}{\underset{H}{C}}—CH_2—\overset{\overset{O}{\|}}{C}—S–ACP$ β-hydroxyacyl-ACP

$\downarrow$ H_2O

$CH_3—\overset{\overset{H}{|}}{C}=\overset{\overset{}{}}{\underset{H}{C}}—\overset{\overset{O}{\|}}{C}—S–ACP$ trans-Δ^2-enoyl-ACP

$\downarrow$ NADPH + H$^+$
$\downarrow$ NADP$^+$

$CH_3—CH_2—CH_2—\overset{\overset{O}{\|}}{C}—S–ACP$ butanoyl-ACP
or fatty-acyl-ACP

Critical Thinking Questions:

1. Work with your team and compare fatty acid synthesis with β-oxidation (CA54A).

   a. Which cofactor is different?

   b. In which pathway is this cofactor synthesized? (Where have we seen it before?)

2. Where in the cell does fatty acid synthesis occur? Why is this important?

3. How many carbons are there in the starting material acetyl-ACP? ____ Malonyl-ACP? ____

4. How many carbons are in the β-ketoacyl-ACP when acetyl-ACP and malonyl-ACP react? Work with your team to describe what happens.

5. Which of the six classes of enzymes (CA 47A) would catalyze each of the elongation reactions shown in Model 1 for fatty acid synthesis? Compare answers with your team.

   a. Step 1 _____

   b. Step 2 _____

   c. Step 3 _____

   d. Step 4 _____

6. During the first spiral of fatty acid synthesis, the four carbon molecule butanoyl-ACP is formed. What happens in the next step to generate a six carbon molecule?

7. Anabolic reactions typically require energy. Consider lauric acid, a 12 carbon molecule.

   a. How many synthesis spirals would be required to make lauric acid? _____

   b. How many malonyl-ACP molecules would be needed? _____

   c. How many NADPH molecules would be needed? _____

   d. If each NADPH is equivalent to 3 ATPs, how many ATP molecules would be required?

   Check to make sure that all team members agree on the answers to parts (a) – (d).

8. A high blood glucose level stimulates the release of the hormone insulin, which activates fatty acid biosynthesis. Discuss with your team why this would be an appropriate response of the organism, and record your explanation in a sentence.

9. Indicate one strength of your teamwork today and why it was beneficial to your learning.

Exercises:

1. Propose some reasons that fatty acid synthesis uses NADPH rather than NADH as a cofactor.

2. Give as many reasons as you can for why fatty acid synthesis cannot be a simple reversal of β-oxidation.

3. Why do fatty acids usually have an even number of carbons?

4. How many spirals of elongation would be required for palmitic acid, $C_{15}H_{31}COOH$? _____

5. Protein catabolism can produce acetyl-CoA. Explain how high protein diets could lead to an increase in body fat percentage.

6. Read the assigned pages in your text, and work the assigned problems.

Amino Acid Metabolism
(Can proteins provide energy?)

Model 1: Overview of amino acid metabolism

Our bodies cannot obtain nitrogen directly from the atmosphere. We must acquire it in our diet, either from plants or animals. Plants obtain nitrogen from the soil for amino acid synthesis, and animals eat the plants for nitrogen. Therefore, the majority of nitrogen is conserved and recycled. Reactions of amino acids are central to this conservation of nitrogen.

Figure 1: Metabolic options for amino acids

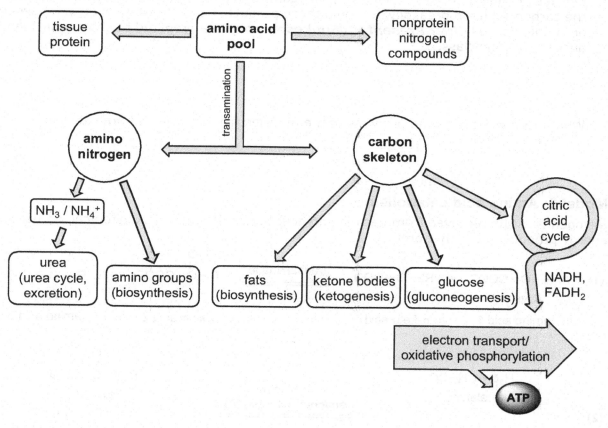

Critical Thinking Questions:

1. What are the three possible options for amino acids obtained either from the diet or protein breakdown? Label them in Figure 1.

2. What is one possible nonprotein nitrogen containing compound that could be made from amino acids? (Hint: consider previous molecules discussed in this course).

3. What are the two general products of transamination?

4. Consider the generic structure for an amino acid below. Discuss with your team what must happen to the structure during *transamination* and write a definition for this term. **Recorder:** write the definition in a complete sentence.

$$R-\overset{\overset{\displaystyle H}{|}}{\underset{\underset{\displaystyle NH_3^+}{|}}{C}}-COO^-$$

5. Which pathway(s) in Model 1 will produce energy?

6. Besides providing energy, work with your team to determine other possible products of the carbon skeleton breakdown. Would these products be made when energy is needed or in abundance? Why? **Presenter (Spokesperson):** Be prepared to share your team's answer with the class.

7. What are the possible end products of the nitrogen from amino acid breakdown?

Model 2: Amino acid catabolism

Nitrogen must be removed from each amino acid before the carbon skeleton can be used for energy production or other products.

(1) $R-\underset{\underset{NH_3^+}{|}}{\overset{\overset{H}{|}}{C}}-COO^- \ + \ R'-\overset{\overset{O}{||}}{C}-COO^- \quad \xrightarrow{\text{transaminase}} \quad R-\overset{\overset{O}{||}}{C}-COO^- \ + \ R'-\underset{\underset{NH_3^+}{|}}{\overset{\overset{H}{|}}{C}}-COO^-$

amino acid 1   α-keto acid 1   α-keto acid 2   amino acid 2

(2)

$\underset{\text{alanine}}{H_3C-\underset{\underset{}{\overset{\overset{NH_3^+}{|}}{CH}}}-COO^-}$

+

$^-OOC-H_2C-H_2C-\overset{\overset{O}{||}}{C}-COO^-$

α-ketoglutarate

$\xrightarrow{\substack{\text{alanine} \\ \text{aminotransferase (ALT)}}}$

$\underset{\text{pyruvate}}{H_3C-\overset{\overset{O}{||}}{C}-COO^-}$

+

$\underset{\text{glutamate}}{^-OOC-H_2C-H_2C-\underset{\underset{NH_3^+}{|}}{CH}-COO^-}$

Critical Thinking Questions:

8. What functional groups are found in the two reactants of reaction (1)?

9. Work with your team to identify what happens to amino acid 1 as it changes to the product α-keto acid 2 in reaction (1). **Recorder:** make note of at least two changes.

10. Reaction (2) is a specific example of transamination.
    a. What products are produced?

    b. What are two possible metabolic routes for pyruvate once it is formed?

    c. What class of molecule is glutamate? Why is this significant?

11. α-Ketoglutarate often functions as an amino group acceptor. Explain why having one predominant acceptor molecule would be favorable in the body.

12. Muscle cells use alanine (in reaction 2) as a "carrier" of nitrogen to the liver. Work with your team to provide a reason the muscle cells would choose alanine and not another amino acid for its nitrogen acceptor. (Hint: consider the molecule used to form alanine. Why might muscles have it in abundance?) **Recorder:** Write your answer in a complete sentence.

Model 3: The role of glutamate

Glutamate serves an important role as a carrier of ammonia to the urea cycle (CA 56).

(3) $^-OOC-H_2C-H_2C-\overset{\overset{\text{H}}{|}}{\underset{\underset{\text{NH}_3^+}{|}}{C}}-COO^-$ + H_2O $\xrightarrow[\substack{\text{glutamate}\\\text{dehydrogenase}}]{\text{NAD}^+ \quad \text{NADH}}$ NH_4^+ + $^-OOC-H_2C-H_2C-\overset{\overset{\text{O}}{||}}{C}-COO^-$

glutamate α-ketoglutarate

Critical Thinking Questions:

13. What reduced coenzyme is formed in reaction (3)? Why would this be significant?

14. What type of reaction is reaction (3)? _____

15. Describe what happens to glutamate in reaction (3) in a sentence.

16. Review Model 1. What is the metabolic option for ammonia (as either NH_3 or NH_4^+) once it is released?

17. Ammonia release (as NH_3 or NH_4^+) from amino acids is a tightly regulated process by which nitrogen passes through a series of intermediates (suchas alanine and glutamate) before release into the urea cycle in the liver. Work with your team to propose a reason that this regulation is essential. **Presenter:** Be prepared to share your answer with the class.

Model 4: Catabolism of Proteins

The carbon "skeleton" of the amino acid left after the removal of the nitrogen can have several fates (Figure 1). When energy is needed, after the nitrogen is removed, the remaining skeleton is targeted for pathways which will produce energy (Figure 2).

Figure 2: Carbons from amino acids are degraded to CAC intermediates and other related metabolites

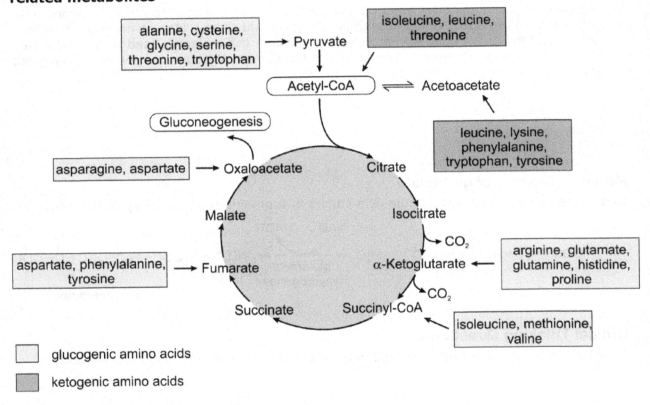

glucogenic amino acids

ketogenic amino acids

Critical Thinking Questions:

18. Assuming that no ATP are used or formed in the transformation of the amino acid serine into pyruvate, how many ATP can be produced using serine for energy? Show your work. Hint: Consider **all** the cofactors produced during its oxidation. Refer to ChemActivity 51 (Citric Acid Cycle), Figure 2, if necessary.

19. Assuming that no ATP are used or formed in the transformation of the amino acid lysine into succinyl-CoA, how many ATP can be produced using lysine for energy? Show your work.

20. Amino acids do not all yield the same amount of energy when they are oxidized. Discuss with your team why this is so. Write your team explanation in a sentence.

21. Write one remaining question for your team about amino acid metabolism.

22. Indicate one strength of your teamwork today and why it helped your learning.

Exercises:

1. Write a definition for a transamination reaction.

2. Why would transamination reactions be referred to as oxidative deamination?

3. Write the products formed when the following amino acids undergo transamination reactions with α-ketoglutarate.
   a. Threonine

   b. Aspartate

   c. Leucine

4. What is the α-keto acid produced in question 3b? In what pathway is this molecule found for generating energy?

5. When the body runs out of glucose, proteins can be broken down to produce glucose and help maintain necessary blood concentrations. Explain how amino acids can be converted into glucose.

6. Read the assigned pages in your text, and work the assigned problems.

(This page is intentionally left blank.)

Urea Cycle
(What happens to excess nitrogen?)

Model 1: Excess nitrogen is transported to the liver (shaded pathway).

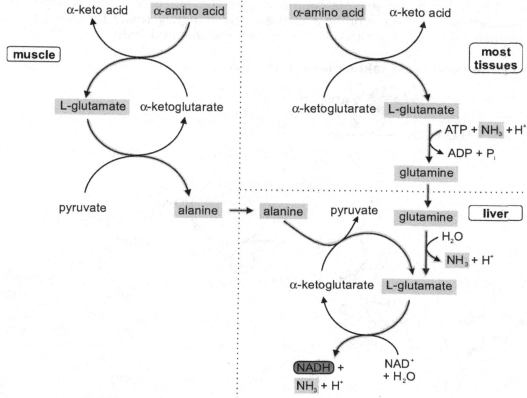

Transport across dotted lines occurs through the bloodstream.

Critical Thinking Questions:

Refer to Model 1 to answer CTQs 1-4.

1. a. When an amino acid is broken down in most tissues, what molecule accepts the nitrogen "waste"?

   b. What happens once glutamate receives nitrogen from an amino acid?

2. a. When an amino acid is broken down in muscle, what molecule accepts the nitrogen "waste"?

   b. What happens in muscle once glutamate receives nitrogen from an amino acid?

   c. Why might muscle cells choose a different molecule to transport the waste nitrogen?

3. Work with your team to trace the pathway of nitrogen starting with an amino acid.

   a. In what organ does the nitrogen end up?

   b. In what molecule does the nitrogen end up?

4. Work with your team to propose one reason why ammonia would be toxic if released into the bloodstream.

Model 2. The urea cycle converts excess ammonia to urea.

Excess ammonia in the liver is converted to urea, which travels to the kidneys to be excreted. The kidneys can excrete ammonia directly, but this requires large amounts of water. Furthermore, excess ammonia in the blood is highly toxic. By converting it to the less-toxic molecule urea, the nitrogen can be safely eliminated from the system.

Figure 1: The urea cycle takes place in the liver.

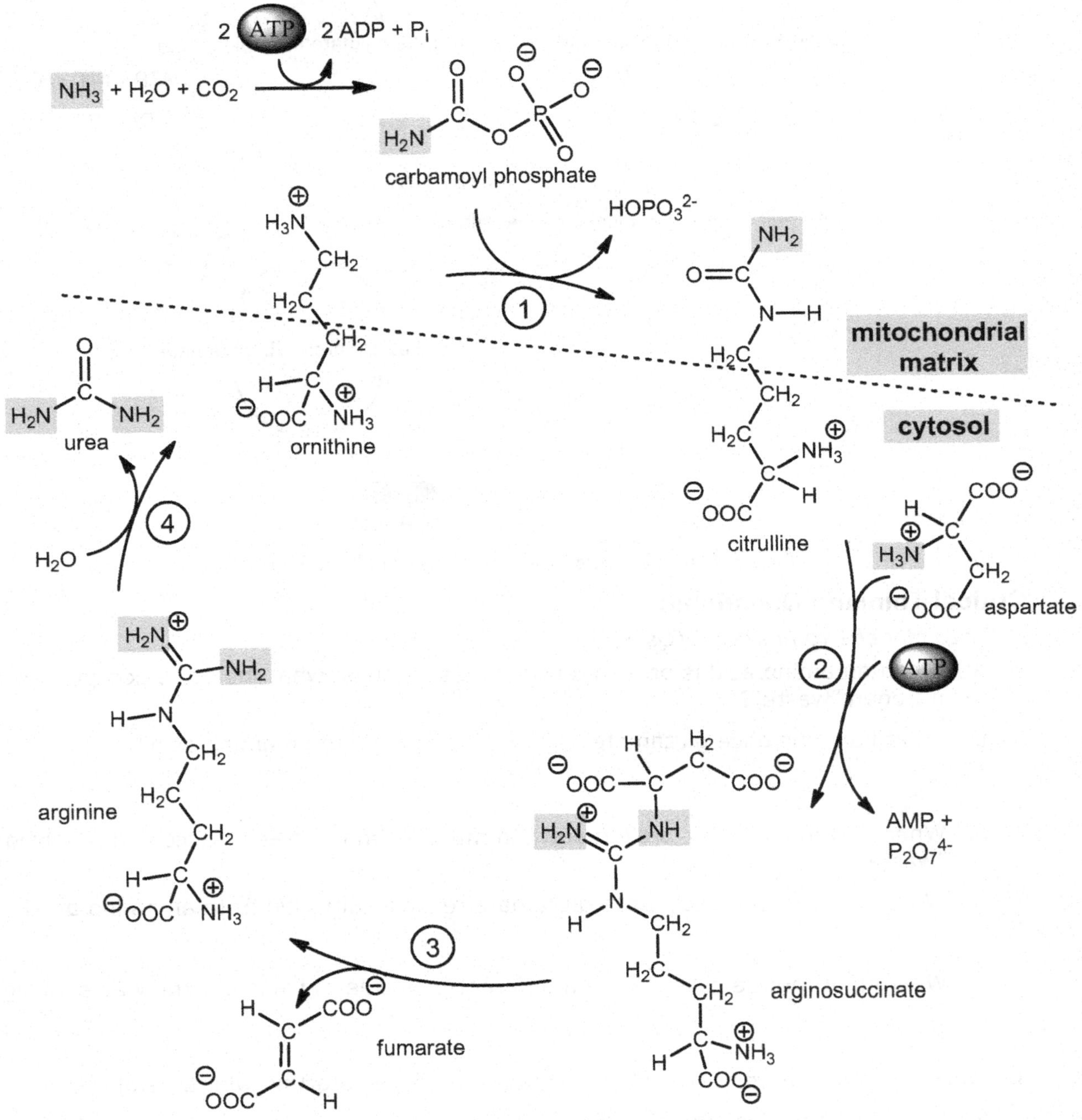

Critical Thinking Questions:

5. Where does the ammonia in Model 2, Figure 1 come from?

6. a. Once the nitrogen is in the mitochondrial matrix, what molecule is formed (Model 2)?

   b. What is the energetic cost for this conversion?

7. a. What product is formed in reaction 1 of the urea cycle?

   b. What must happen to this product before reaction 2 of the urea cycle?

8. Review reactions 2 and 3 of the urea cycle with your team (from citrulline to arginine). Draw a circle around the part of the arginine molecule that is contributed by aspartate.

9. Discuss with your team to determine the significance of the portion aspartate contributes to the cycle. Write your consensus answer.

10. a. What product is "released" from the cycle in reaction 3?

    b. In which other pathway is the product from (a) found?

11. a. What class of organic reaction cleaves urea from arginine in step 4 of the urea cycle?

    b. What other product is formed? Why is this product important?

12. How many total ATPs are used in this process to make urea? How many high energy bonds?

13. Urea eliminates two nitrogens from the body. Work with your team to determine the molecule(s) in Figure 1 from which each nitrogen originates.

14. What questions remain with your team about the elimination of excess nitrogen from the body?

15. Describe one strength of your teamwork today and how it contributed to your understanding of the material.

Exercises:

1. Write the net reaction for the urea cycle.

2. Urea contains one carbon. From which molecule does the carbon originate?

3. Several diseases result from a defect in one of the urea cycle enzymes. Propose an adjustment in diet that might help minimize the need for the urea cycle.

4. Considering pathways we have discussed previously (Citric Acid Cycle, CA51 and Amino Acid Metabolism, CA55), propose a pathway for fumarate to generate more aspartate.

5. Read the assigned pages in your text, and work the assigned problems.

Periodic Table of the Elements*
with average atomic masses to two decimal places

1	2	3	4	5	6	7	8	9	10	11	12	13	14	15	16	17	18
1 **H** 1.008																	2 **He** 4.003
3 **Li** 6.941	4 **Be** 9.012											5 **B** 10.81	6 **C** 12.01	7 **N** 14.01	8 **O** 16.00	9 **F** 19.00	10 **Ne** 20.18
11 **Na** 22.99	12 **Mg** 24.31											13 **Al** 26.98	14 **Si** 28.09	15 **P** 30.97	16 **S** 32.07	17 **Cl** 35.45	18 **Ar** 39.95
19 **K** 39.10	20 **Ca** 40.08	21 **Sc** 44.96	22 **Ti** 47.88	23 **V** 50.94	24 **Cr** 52.00	25 **Mn** 54.94	26 **Fe** 55.85	27 **Co** 58.93	28 **Ni** 58.69	29 **Cu** 63.55	30 **Zn** 65.39	31 **Ga** 69.72	32 **Ge** 72.61	33 **As** 74.92	34 **Se** 78.96	35 **Br** 79.90	36 **Kr** 83.80
37 **Rb** 85.47	38 **Sr** 87.62	39 **Y** 88.91	40 **Zr** 91.22	41 **Nb** 92.91	42 **Mo** 95.94	43 **Tc** (98)	44 **Ru** 101.1	45 **Rh** 102.9	46 **Pd** 106.4	47 **Ag** 107.9	48 **Cd** 112.4	49 **In** 114.8	50 **Sn** 118.7	51 **Sb** 121.8	52 **Te** 127.6	53 **I** 126.9	54 **Xe** 131.3
55 **Cs** 132.9	56 **Ba** 137.3	57 **La** 138.9	72 **Hf** 178.5	73 **Ta** 180.9	74 **W** 183.9	75 **Re** 186.2	76 **Os** 190.2	77 **Ir** 192.2	78 **Pt** 195.1	79 **Au** 197.0	80 **Hg** 200.6	81 **Tl** 204.4	82 **Pb** 207.2	83 **Bi** 209.0	84 **Po** (209)	85 **At** (210)	86 **Rn** (222)
87 **Fr** (223)	88 **Ra** 226.0	89 **Ac** 227.0	104 **Rf** (265)	105 **Db** (268)	106 **Sg** (271)	107 **Bh** (272)	108 **Hs** (277)	109 **Mt** (276)	110 **Ds** (281)	111 **Rg** (280)	112 **Cn** (285)	113 **Nh** (284)	114 **Fl** (289)	115 **Mc** (288)	116 **Lv** (293)	117 **Ts** (292)	118 **Og** (294)

58 **Ce** 140.1	59 **Pr** 140.9	60 **Nd** 144.2	61 **Pm** (145)	62 **Sm** 150.4	63 **Eu** 152.0	64 **Gd** 157.3	65 **Tb** 158.9	66 **Dy** 162.5	67 **Ho** 164.9	68 **Er** 167.3	69 **Tm** 168.9	70 **Yb** 173.0	71 **Lu** 175.0
90 **Th** 232.0	91 **Pa** 231.0	92 **U** 238.0	93 **Np** 237.0	94 **Pu** (244)	95 **Am** (243)	96 **Cm** (247)	97 **Bk** (247)	98 **Cf** (251)	99 **Es** (252)	100 **Fm** (257)	101 **Md** (258)	102 **No** (259)	103 **Lr** (262)